SOLAR POWER

THE BEGINNER'S GUIDE FOR SOLAR ELECTRICITY SYSTEMS, FOR OFF-GRID SOLAR SYSTEMS AND FOR YOUR HOME ENERGY INDEPENDENCE

Turgon Annárë

©Copyright 2020 by Turgon Annárë

All rights reserved.

This document is geared towards providing exact and reliable information with regards to the topic and issue covered. The publication is sold with the idea that the publisher is not required to render accounting, officially permitted, or otherwise, qualified services. If advice is necessary, legal or professional, a practiced individual in the profession should be ordered.

-From a Declaration of Principles which was accepted and approved equally by a Committee of the American Bar Association and a Committee of Publishers and Associations.

In no way is it legal to reproduce, duplicate, or transmit any part of this document in either electronic means or in printed format. Recording of this publication is strictly prohibited and any storage of this document is not allowed unless with written permission from the publisher. All rights reserved.

The information provided herein is stated to be truthful and consistent, in that any liability, in terms of inattention or otherwise, by any usage or abuse of any policies, processes, or directions contained within is the solitary and utter responsibility of the recipient reader. Under no circumstances will any legal responsibility or blame be held against the publisher for any reparation, damages, or monetary loss due to the information herein, either directly or indirectly.

Respective authors own all copyrights not held by the publisher.

The information herein is offered for informational purposes solely, and is universal as so. The presentation of the information is without contract or any type of guarantee assurance.

The trademarks that are used are without any consent, and the publication of the trademark is without permission or backing by the trademark owner. All trademarks and brands within this book are for clarifying purposes only and are the owned by the owners themselves, not affiliated with this document

CONTENTS

PREFACE -- **1**

INTRODUCTION -- **14**

CHAPTER ONE --- **16**

   WHAT IS SOLAR ENERGY --------------------------- 16

CHAPTER TWO -- **31**

   TYPES OF SOLAR -- 31

CHAPTER THREE --------------------------------------- **52**

   PHOTOVOLTAIC SYSTEM ------------------------------ 52

CHAPTER FOUR --- **62**

   PASSIVE SOLAR SYSTEM ----------------------------- 62

CHAPTER FIVE -- **69**

   CONSERVATION RULES ------------------------------ 69

CHAPTER SIX -- **77**

   OFF-GRID SOLAR PV SYSTEM --------------------- 77

CHAPTER SEVEN --------------------------------------- **91**

ABOUT INSTALLING PANELS AT HOME ------- 91

CHAPTER EIGHT --------------------------------------- **119**

TIPS FOR USERS OF PHOTOVOLTAIC SOLAR ENERGY INSTALLATIONS: ----------------------------- 119

CHAPTER NINE -- **147**

WHAT IS A POWER SUPPLY ------------------------- 147

CHAPTER TEN --- **154**

SOLAR ENERGY FOR YOUR HOME ------------- 154

CHAPTER ELEVEN ------------------------------------ **162**

COMMON MISTAKES WHEN INSTALLING SOLAR PANELS -- 162

FAQ FREQUENT QUESTIONS ---------------------- **168**

CONCLUSION --- **179**

PREFACE

START YOUR SOLAR JOURNEY

Do you want more control over your energy use? Do you want to reduce reliance on traditional generators, distributors and retailers? There are several steps along the path to complete energy independence.

The go to for energy independence is the solar energy. The easiest and most effective way to gain energy independence is to install solar power. Generate your own power and export the surplus to the grid. With benefits to include three years maintenance years, there is no need to check or worry. All you have to do is enjoy reduced electricity rates.

Some appliances (such as lights) need to run at night, while others (dishwashers, washing machines, etc.) can be turned on at selected times.

Waiting for the sun to shine, the solar panel can run the entire cycle using only solar energy. Even assuming that it is not optimal and not a completely cloudy day, the solar panel system will still generate enough power.

This is particularly attractive if you can only get a lower price for excess solar power (e.g. 8 c / kWh), but the higher retail prices for electricity used at night (e.g. 26 c / kWh)).

To ensure enough solar power to run the appliance at home, install a home display that links to a low-cost power meter or smart meter. These devices can display the amount of power exported to the grid. Some brands of inverters can be fitted with a relay output that will notify you when it is generating excessive solar energy.

Increase energy efficiency

Energy efficiency does not mean returning to the Dark Age. This is only when using energy in a smarter way. Energy-efficient households are households that do not require as large a photovoltaic system or energy storage system.

There are many ways to reduce energy consumption, but there are quite a few that are completely inexpensive to implement.

Household energy storage – "light"

In the day, energy is generated into the solar system wich will be put to use most of their energy after work and on weekday nights. This discrepancy can be resolved by "time shifting" the solar using a small battery bank.

A home energy storage system means that much of the solar energy it generates stays at home, and less electricity is exported to the grid. It also means that you can avoid buying electricity from the grid during peak times. From 4pm to 10pm. This is especially attractive if you use hours of use when your electricity rates fluctuate during the day.

The small battery bank is fully charged every day and runs out every night.

For each unit of solar energy that can be stored in a battery for night use, it saves the cost difference between a fixed price (eg 8 c / kW · h) and a retail price (eg 26 c / kW · h).

Household energy storage – "robust"

Larger battery systems are needed to maximize energy independence, self-sufficiency and the amount of solar energy used at home. The batteries need not be as large or expensive as a true "off-grid" system. This is because the grid can be used for long periods of winter or bad weather. This means that the battery capacity can be 1-2 days instead of the 3-4 days that off-grid system designs can handle.

At present, technology is evolving rapidly, even when a complete battery system is out of reach. Prices should drop significantly, as demonstrated by solar panels.

Until then – install solar and take the first step.

You can work through these steps of the design process and get to know your system. Right here, right now.

Easy Steps to Design Your Own System

The first step is to assess your current energy usage. In order to design the right system for you, you need to know how much energy you are currently using. If you want to save money on your system, consider ways to reduce energy usage.

The second step is to look at the various systems available and better understand which type of system will meet your needs.

Once you know the type of system, the next step is to figure out the size of the system. If you plan to live off-grid, you need to know how much power your appliances consume. Maintaining a connection to the grid requires the required average kilowatt hours (kWh).

The Types Of Solar Systems?

System Size

The grid tie system connects to an electric grid. The power generated by the solar system offsets the power consumed. To determine the size of the Grid Tie system, you need to know your average monthly kilowatt-hour (kWh) usage. This can be found on your monthly invoice. System sizing can also be based on the space available on the roof or parcel. All systems can be added over time, so if your budget is a factor, you can start small and add as much as your budget allows.

Panel

One of the first things to consider is what kind of panel to use and how many pieces will be needed. Also to be put in consideration is which location on the roof these panels will occupy.

Inverter

Solar panels produce direct current (DC) power, but homes use alternating current (AC) power. An inverter is required to convert power from DC to AC so that it can be used. Grid Tie inverters have three options: centralized string inverters, micro inverters and solar edge inverters.

Racking

Solar panels can be mounted on the roof or on the ground. The exact cost of the mounting system depends on the roofing or ground-mounted configuration. The average price for a roof mount option is $ 50 per panel, and the average price for a ground mount option is $ 80 per panel.

Battery backup system

Gridtie PV systems generally only work when commercial power is available. When the grid goes down, the electricity from the solar panels is cut off. Currently, Four Star Solar's battery backup system with its special power center allows grid-connected solar panels to charge a battery bank that can power the house.

EVALUATE YOUR ENERGY USE

The first step in designing a new renewable energy system is to assess current power usage.

Start by identifying your kilowatt-hour (kWh) usage for 12 consecutive months. Check with your utility to make sure this information is available online, call, or look at past electricity rates.

Sum the kWh for each month to find the average kWh usage and divide it by 12. Make note of the month that uses the most energy.

Make a note of this number.

The average kWh per month helps determine the size of the PV system.

If cost is not an issue or you are already saving energy with a sustainable green lifestyle, you can skip the information below and go to "Step 2: Determine the type of system you need" I can do it.

There is a monthly average kWh.

And to step two

DETERMINE WHAT SYSTEM

For any dollar expense on energy efficiency, you can save up to $ 5 on solar system costs.

Reduce current energy usage

A further assessment of current energy usage reveals exactly where energy and cost efficiencies can be achieved and ultimately maximizes the use of solar PV systems.

A convenient and easy way to see the exact location where you can reduce costs and reduce usage is to use an inexpensive and sophisticated home energy monitoring device. Use a small meter installed on your home energy panel to send data to your Wi-Fi network and combine it with the included monitoring app to get detailed energy usage data wherever you are.

Utilities and municipal programs are the perfect resource for energy saving plans (and cost reduction incentives). They may have free or low cost programs to assess your energy use, and your home may lose energy through ducts, walls, windows, etc.

You can take advantage of a "smart home" computer program or company. Some programs are set to alert you when a home appliance or zone is using more power than necessary. These energy management techniques can be as simple as using a simple, cost-effective app.

HOW TO SAVE ENERGY

Reduce phantom load

Phantom load refers to the continued draw of electricity when the appliance is in the "off" or "standby" position. They can consume 24 hours a day power. Also, some appliances approach full power just to be on standby. Some common examples are:

- computer
- stereo
- TV set
- VCR
- gas oven glover

- Electronic phone

- A power cord with a small "box"

An easy solution is to connect to a power strip. Turn off the power strip when you're not using items.

Energy efficient lighting

There are other things you can do besides changing all the bulbs. Use sunlight for reading, working and living. Low watt task lighting can replace high energy common overhead. The bright colors of the walls reflect more light, and solar tubes or skylights can further improve dark areas.

The new high efficiency standard for light bulbs was enacted by the Energy Independent Security Act (EISA) 2007 and became effective nationwide on January 1, 2012.

The new Energy Star label shows lumens, estimated annual cost, life expectancy, color, and watts. Look at the required brightness lumens and the wattage of energy used. This ultimately makes it easier to choose the type of bulb.

Traditional 100-watt incandescent bulbs have been replaced by 72-watt incandescent bulbs that emit about

1,600 to 1,700 lumens. Compact fluorescent lamps that emit the same lumen use only 23 watts. More efficient incandescent bulbs are cheaper, but the energy savings of compact fluorescent or LED bulbs are higher than the higher initial costs.

Compact fluorescent lamps typically save $ 3 per year per bulb replaced (when used 4-6 hours a day). CFL bulbs can go up to ten times longer than traditional incandescent bulbs.

LED (Light Emitting Diode) bulbs provide the best colors, at least as good as incandescent bulbs LED bulbs are more durable and do not break as easily as incandescent or CFL bulbs. LEDs are initially more expensive than CFLs, but last up to five times longer than CFLs.

Both CFLs and LEDs use about 75% less energy than incandescent bulbs and do not sacrifice light. Also, almost no heat is generated. For the best warranty and long-lasting lights, look for an Energy Star rated bulb. For example, after about a year, poor quality LEDs may be dark and uneven, flicker, change color, or continue to use power when turned off.

HOW TO REDUCE HEATING COSTS

1. Insulation-One of the most cost-effective ways to make your home more energy efficient and comfortable

is to add insulation. A poorly insulated attic can be the largest source of energy loss as heat passes through the ceiling and enters and exits the attic. Walls are the second place where heat escapes, especially in older houses where the insulation is worn or settled, or simply un-insulated. Insulated window curtains are good as it help trap heat at night. Remember that even double-pane windows have little insulation value.

2. Seal air leaks. Sealing air leaks to block out cold air, such as adding windproof windows or caulking the outside of windows, prevents drafts from entering the insulated house.

3. Use a setback thermostat. These can be installed and programmed to reduce household temperatures as needed. The thermostat can also be set back manually. By reducing the room temperature by 5 degrees for 8 to 12 hours, you can save 5% on heating costs.

4. Use passive solar technology. Passive solar technology captures the energy of the sun to warm the house and avoids capturing the same energy when you need to keep the house cool. House designs that take into account changes in the angle of the sun from winter to summer can dramatically lower energy costs.

Water heating cost at the bottom of the insulated pump

Hot water is a major energy cost in homes. It usually accounts for 13% of the utility bill. There are six ways to reduce hot water energy consumption.

1. Insulate the water heater. Save money by wrapping a gas or electric water heater in an insulated jacket. Wraps are easily available in hardware stores and can be installed by homeowners.

2. Lower the thermostat of the water heater. To optimize efficiency, lower the thermostat to 110 degrees Fahrenheit. With an electric water heater, attaching a timer to adjust the heating cycle

3. Insulate the pipe. Too much heat is lost from the pipe coming directly from the hot water heater. With more lost in un-heated crawl spaces. Heat insulation in these areas significantly reduces heat loss in these areas. Users use little time waiting for hot water at the tap and less waste. On-demand hot water circulation is an exciting innovation in this area and can maximize efficiency.

4. Attach the aerator to the faucet. These reduce the flow of hot and cold water while maintaining the original water pressure.

5. Install low flow shower head. These reduce water usage by up to 50%.

6. Reduce the amount of hot water used. Fifteen percent of the energy used to heat water in the average

home can be significantly reduced by using simple formulas for insulation and savings. Save money by maximizing home efficiency.

Renewable energy systems are in many sizes and shapes. Some have solar panels. Others are due to wind, water, or a combination of these. Some are grid connected. Other utilities are independent of power lines.

What are the main reasons to add a solar: save power or be independent of the power company?

Off-grid systems allow you to work with electricity in areas where public power is unavailable or too expensive to bring in. Save money

Solar energy remains one of the best and comfortable ways to power our homes. Despite the inconveniences, the use of solar energy has increased by about 20 percent per year in the last 15 years, thanks to the rapid fall in prices and the increase in efficiency.

INTRODUCTION

The name of solar energy that which comes from the direct use of the sun's radiation, and from which heat and electricity are obtained, heat is obtained through thermal collectors, and electricity through photovoltaic panels.

In the systems of thermal exploitation, the heat collected in the solar collectors can be used to meet numerous needs, such as: obtaining hot water for domestic or industrial consumption, or for heating purposes, agricultural applications, among others.

The photovoltaic panels, which consist of a set of solar cells, are used for the production of electricity, and constitute an adequate solution for the electricity supply in rural areas that have an abundant solar resource. The electricity obtained through photovoltaic systems can be used directly, or stored in batteries for overnight use.

The simplest method for solar capture is that of photovoltaic conversion, which consists of converting solar energy into electrical energy by means of solar cells. These cells are made of pure silicon with the addition of impurities of certain chemical elements, and are capable of generating each 2 to 4 Amps, at a voltage of 0.46 to 0.48 V, using radiation as raw material solar. They admit both direct and diffuse radiation, which means that you can get electricity even on cloudy days.

The cells are mounted in series on solar panels or modules to achieve a voltage suitable for electrical applications; the panels capture solar energy by transforming it directly into electricity in the form of direct current, which must be stored in accumulators

The era of renewable energy is here to stay. Learn what solar panels are, how they work and what are the benefits and limitations of this technology.

Knowing how solar panels work is the first step to get fully into the universe of solar energy and explore ways to spend less money, while taking care of the environment.

Solar energy will be the vendetta of the new era. Produced by light (photovoltaic energy) or by the heat of the sun (solar thermal), it can generate electricity or produce heat and is one of the technologies that is closer to displacing oil from its throne.

It is that, with the constant increase in the demand for traditional energy sources and the consequent increase in costs, solar energy is increasingly a necessity.

Therefore, the use of the inexhaustible source of energy that the sun provides us through the use of solar panels or solar hot water tanks becomes an excellent alternative both domestic and industrial.

CHAPTER ONE

WHAT IS SOLAR ENERGY

Solar energy, also called solar energy, refers to the energy of solar radiation, which can be used by humans technically. The use can take place in the form of electricity, as heat, but also as chemical energy.

Solar energy is produced by light - photovoltaic energy - or heat from the sun - thermo solar - for electricity generation or heat production. Inexhaustible and renewable, since it comes from the sun, it is obtained through panels and mirrors.

The photovoltaic solar cells convert sunlight directly into electricity by photoelectric effect called, whereby certain materials are able to absorb photons (light particles) and release electrons, generating an electric current.

The term solar energy refers to the production of thermal and electrical energy obtained through the use of the sun's rays. The sun radiates our planet at a power of about 180 billion kilowatts. A part of the sunlight is reflected by the atmosphere towards outer space.

At any time the sun radiates to the Earth's orbit an energy equal to 1367 watts / m² (1.3 kW / m 2). In general, it reaches the Earth's surface about 1 kilowatt of solar energy per square meter. The energy produced by solar rays can be harnessed by the use of different renewable technologies, such as solar panels.

Solar energy is one of the main sources of renewable energy.

On the other hand, solar thermal collectors use panels or mirrors to absorb and concentrate solar heat, transfer it to a fluid and conduct it through pipes for use in buildings and facilities or also for the production of electricity (thermoelectric solar).

This is only possible through the nuclear fusion processes in the interior of the sun. By burning hydrogen on its surface prevailing temperatures of about 5500 degrees Celsius, which meet as electromagnetic radiation partly on the earth's surface.

LITTLE SOLAR ENERGY REACHES THE EARTH'S SURFACE

While solar energy at the boundary to the earth's atmosphere has an intensity of 1.367 kW per square meter, the so-called solar constant, the radiation energy on the earth's surface hits "only" 0.114 to 0.268 kW per square meter, depending on the region.

A large part of the incident solar energy is already absorbed or reflected by the earth in advance. Nevertheless, this "low" intensity on the earth's surface is sufficient for man to use it technically in different areas of the energy supply.

THE BIGGEST SOURCE OF ENERGY FOR HUMANITY

Since solar energy can be measured at all, the sun emits an almost constant radiant energy. Serious fluctuations are not known even from times long past. Thus, solar energy represents an almost limitless source of energy that, unlike fossil fuels, is hardly exhaustible.

The sun, or solar energy, is therefore the largest, available energy source of humanity and with modern solar technology in different energy ranges for humans usable.

SUNSHINE WORLDWIDE

Solar energy can be used for example by means of solar collectors for heat generation, generated within solar thermal power plants with the aid of steam electricity or is used for DC generation by photovoltaic systems. However, solar energy or solar radiation is subject to daily, seasonal and regional fluctuations.

Theoretically, a solar farm covering an area of 700 x 700 kilometers, set up in the sun-drenched Sahara and with an efficiency of only 10 percent, would be enough to cover the global total energy demand.

SOLAR ENERGY EVEN IN SUN-POOR AREAS

In order to be able to use the solar energy independently of the radiation intensity, for example during the night hours, the solar energy can be stored. However, storage is one of the biggest challenges and is driven by targeted research.

Also a possible improvement in the efficiency of the technical components of solar collectors and especially photovoltaic systems in areas with less solar energy is the focus of research. A development that makes it possible to use the available solar radiation as effectively and inexpensively as possible

SOLAR ENERGY: MAIN TECHNOLOGIES

PHOTOVOLTAIC SOLAR PANEL

The photovoltaic solar panel directly converts the sun's energy into electricity through the use of the physical properties of some semiconductors when they are stimulated by sunlight.

SOLAR THERMAL PANEL

The solar thermal panel (or solar collector) is a technology capable of capturing the thermal energy of the solar rays to heat the hot water contained in a storage tank through a heat exchanger.

The concentration of the solar panel captures the sun's rays through a system of parabolic mirrors with a linear structure that focus towards a single point where a heat transfer fluid flows or to a boiler. Solar energy is huge, it is considered a renewable and inexhaustible energy source on the human time scale.

Unlike fossil energy sources, solar energy is considered inexhaustible, since it is based on the concept of flow rather than values. The exploitation of solar energy does not reduce its future availability in terms of flow. However, solar energy is also an intermittent source (day / night) and does not concentrate two characteristics that are an obstacle to large-scale exploitation.

Currently solar energy is mainly used to produce hot water (solar thermal energy) and to produce electricity (photovoltaic). The first medium-sized solar power plants to produce electricity are also underway. Solar panels and solar energy are also used in the aerospace industry to provide electricity to satellites, spacecraft stations or space. In the future, the solar orbital center to collect the sun's rays directly in space and transmit energy to the earth's surface.

SOLAR FURNACE

The sun's energy is the "mother" source of all energy sources on Earth and is a primary energy source. Directly or indirectly, all sources of energy derived from solar activity and life itself would not exist on our planet. Just think of wind energy.

Without the sun there would be no continuous movement of the air masses and the energy of the wind could not be harnessed. There would be no life and, therefore, not even fossil fuels; there would be no rain (hydroelectric), vegetables (biomass) and so on. Solar radiation has created the ideal conditions for plant life through photosynthesis. Without the energy of the fossil sun stored in coal, oil and gas, man could not have entered the process of the industrial revolution of his own society.

HISTORY OF SOLAR POWER

Within the history of solar energy, one way or another, solar energy has always been present in the life of the planet, which is essential for the development of life. However, the way in which human civilization has taken advantage of it invented new strategies and tools have undergone a long evolution.

The Sun is indispensable for the existence of life on the planet: it is responsible for the water cycle, photosynthesis, etc. Already the first civilizations realized this and, as civilizations have evolved, techniques to harness their energy have also evolved.

At first they were techniques to take advantage of passive solar energy, later techniques were developed to take advantage of thermal solar energy , and subsequently photovoltaic solar energy was added .

THE SUN AND THE ANCIENT CIVILIZATIONS

Architecture of the ancient civilizations dedicated to the Sun throughout the history of solar energy, the Sun has always been an essential element for the development of life. The most primitive cultures have been taken advantage of indirectly and without being aware of it.

Later, many more advanced civilizations realized the importance of the Sun and developed numerous religions that revolved around the solar star. In many cases, the architecture also kept a close relationship with the sun . Examples of these civilizations would be found in Greece, Egypt, the Inca Empire, Mesopotamia, the Aztec Empire, etc.

PASSIVE SOLAR ENERGY

In the aspect of passive solar energy, it is worth mentioning the role of the Greeks who were the first in history to design their houses to take advantage of sunlight , probably since 400 BC

Another important moment in the history of solar energy was the Roman era. During the Roman Empire, glass was first used in windows to harness light and trap solar heat in their homes. They even enacted laws that penalized blocking access to light to neighbors.

The Romans were the first to build glass houses or greenhouses to create suitable conditions for the growth of exotic plants or seeds that brought Rome from the far reaches of the empire.

HISTORY OF SOLAR THERMAL ENERGY

It is quite difficult to specify a specific date when contacts between human beings and solar energy began. Activities as normal as drying clothes taking advantage of the sun's heat date back to prehistoric times

But when did you start using tools to improve that use? Experts base their origins between the seventh and seventh centuries BC. C., when the Roman priestesses lit the fires of the temples thanks to a system of concave mirrors used to reflect sunlight.

In the military field it has also been used. The first of them known was a great Greek physicist, Archimedes, who in 212 a. C. invented a machine that concentrated the sun's rays. Legend has it that with it he managed to burn some Roman ships to repel an attack on Syracuse.

Subsequently, a genius like Leonardo Da Vinci tried to invent in 1515 a six kilometer diameter heat concentrator, although he never managed to finish it and a hundred years later, the French Salomon de Caux developed a water pump to run a small fountain thanks to the expansion of a metal container due to the heat of the sun.

Focusing on the most important advances for the development of solar thermal energy as we know it today, we must highlight the work of Horacle de Saussure in 1767. This Swiss wanted to check what temperature could be reached in a glass box thanks to the greenhouse effect.

He managed to reach 109 degrees Celsius and unknowingly created the so-called solar collector. This invention served for the development of water heating systems by the sun's rays. In addition, he was also in charge of inventing the first solar oven to prepare meals.

In 1891 it was Clarence Kemp who devised the first solar-powered water heater thanks to the use of a collector. Registered the patent as 'Climax', boosting the commercial development of this type of energy

MOTIVATION FOR GOING SOLAR

Solar energy is the energy whose source is from the sun. It is the cleanest type of energy source. It is always available and cannot be exhausted.

Because the sun is always a source of energy, its energy is renewable and you don't have to worry about the light going out.

Another advantage of solar energy is that, unlike fossil fuels that emit greenhouse gases, solar cells do not emit anything into the air and are therefore environmentally friendly.

Solar energy operates in a silent system, so there is no noise pollution.

ADVANTAGES OF SOLAR POWER

The advantage of solar is numerous, as it is always available and cannot be finished. It is not regulated by any authority like the case of oil or other mineral resources.

The energy produced by the sun has its advantages and disadvantages. The advantages are mainly concentrated in the lower environmental impact and the possibility of exploitation of the perennial source of solar energy.

- It saves fuel costs and becomes more independent of the rising energy prices due to the depletion of oil and gas reserves.
- If you invest today, you will immediately get solar energy for free.
- With solar technology, you take home one of the most modern water heating systems. With a technique that meets the provisions and requirements of current regulations and anticipates future ones.
- The Public Administrations are committed to solar technology, and for this they have provided various support programs, with economic incentives that help them carry out their solar installation.
- It protects the environment and shows orientation towards the future and towards the new energies that are already present.
- You are free from the noise of generator or the stress of putting it on.

SUMMARY OF SOLAR ENERGY BENEFITS

- Renewable
- Inexhaustible
- Non-polluting
- Avoid global warming
- Reduce the use of fossil fuels
- Reduce energy imports
- Generate wealth and local employment
- Contributes to sustainable development
- It is modular and very versatile, adaptable to different situations
- Allows applications for large-scale power generation and also for small cores isolated from the network

DRAWBACKS TO SOLAR POWER

The disadvantages of solar energy are the following:

SCHEDULE LIMIT

Keep in mind that the sun is hidden at sunset, so you could only enjoy this fountain during the day and not at night. We also have to know that in certain areas or countries the solar schedule is much shorter than in other parts.

IT IS NOT SUITABLE FOR WINTER

In some countries when winter comes the hours of sunshine are very few. Because the solar inclination varies constantly, this will totally interfere with the performance of solar panels, making them very inefficient.

LIMITATION FOR ENERGY STORAGE

Because the hours where the energy can be obtained, will not always coincide with the hours to store this, it will not always be possible to have energy reserves for when necessary.

IT DEPENDS ON THE WEATHER

The performance of solar energy will depend entirely on the weather function. So this energy will be very unfeasible in areas where the sky is mostly cloudy.

IT GENERATES HIGHER EXPENSES

Since the production of solar energy will depend on the climatic changes, it is necessary to have other sources of energy or fuels to complete its production.

GREAT INVESTMENT

Due to the high cost of solar panels, a strong initial investment must be made. Its maintenance, although it is recommended to do it annually, is very expensive.

It has been proven that it will take 10 to 15 years to recover only the cost of the initial investment.

HIS APPEARANCE IS NOT PLEASANT

The shape of the solar panels is not in good taste and its appearance is not very pleasant to look at.

IT PRODUCES A GREAT ENVIRONMENTAL IMPACT

Solar panels are manufactured with certain molecules, such as lead or silicon, that must be treated as highly hazardous waste at the end of their useful life.

It will also take several years for the degradation of these materials.

INTERMITTENT ENERGY

As we had already mentioned, sunlight will depend on solar schedules. Therefore, access to this energy will be limited to certain times of the day and its production will be impossible on cloudy or rainy days.

IT IS NOT FEASIBLE FOR THE WORLD POPULATION

It has been shown that it is not possible to feed the planet solely and exclusively with solar energy, it would only work if there were large batteries that have the capacity to store large amounts of energy, and then gradually release it when necessary for the general population.

Although there are more advantages than disadvantages generated by the production of solar energy, it is verified annually that the inconveniences caused by said energy have generated a great environmental impact and would be contributing negative aspects to the planet earth.

- The economic cost compared to other options.
- The performance is a function of the weather.
- Limitations on solar hours.
- Limitations to store the generated energy
- In addition, the facility needs proper regular maintenance to function properly.
- There is no repairing in the major parts of the solar system, it usually needs to be replaced when faulty.

Finally, in many cases, the areas with the highest solar radiation are desert areas far away from the consumption areas.

CHAPTER TWO

TYPES OF SOLAR

There are two types of use of solar energy: the one used to produce thermal energy (basically, domestic hot water and heating) and the one that converts solar radiation into electricity using the so-called photovoltaic technology.

The possibility of making more global use of solar radiation in building construction must also be considered. This way of harnessing solar energy is called bioclimatic architecture and takes into account the natural light and weather conditions of each site for the construction of new homes.

The solar thermal installations consist of a system for capturing the radiation that comes from the sun (the solar collector), a storage system for the thermal energy obtained (the storage tank) and a system for distributing heat and consumption.

The most widespread and known applications are those of low temperature, that is, those that provide heat at a temperature below 100oC. The main components of this type of installation are described below.

THERMAL SOLAR ENERGY

Solar thermal collector

The flat glazed roof sensor is the type of sensor that, until now, has had more diffusion. Its uses the greenhouse effect in operating, that is, it captures the solar radiation inside it, transforms it into thermal energy and prevents the exit to the outside.

The main elements that make up a solar collector with glazed roof are:

- Transparent cover
- Absorbent surface
- Circulation tubes
- Isolating material

Types of solar energy solar radiation, upon reaching the collector, crosses the transparent cover and affects the absorbent surface, which captures this radiation and transmits it, in the form of thermal energy, to the circulating fluid. Normally, this fluid is water with an antifreeze liquid, although it can also be air, in the so-called air collectors, which are normally used for heating.

As a general rule, the collectors have to be installed facing south to capture the maximum solar radiation, and their inclination with respect to the horizontal plane has to be equal to the latitude of the site.

Accumulator tank

It serves to accumulate energy at times of the day when it is possible and use it when demand occurs. In small installations, it is possible to incorporate the accumulator in the upper part of the collector: they are the equipment called thermosiphons, which take advantage of the water circulation due to temperature differences (convection).

Heat and consumption distribution system

It consists of a control and management system for facilities, pipes and ducts, pumps to circulate fluids, air traps and various valves.

Support system

To replace possible periods without sunshine, solar thermal installations incorporate a conventional water heating system, which is only used when the energy received in the collectors is not enough.

Solar thermal installations can be made as open circuits or as closed circuits, depending on whether the same water that circulates through the solar collectors is drinking water or not.

Open circuit installations are simpler, but they have the disadvantage of the danger of frost, corrosion or encrustation in the collector. In closed circuit systems there is no mixing between the liquid that circulates through the collectors (primary circuit), to which an antifreeze is added, and the water destined for consumption (secondary circuit).

PHOTOVOLTAIC SOLAR ENERGY

Photovoltaic solar energy is the transformation of light into electrical energy through solar cells, providing economic and environmental benefits.

It is an inexhaustible energy, clean, silent and respectful of the environment. In addition, it is available anywhere.

Solar cell

The solar cell is a semiconductor where, artificially, a permanent electric field has been created, whereby, when the solar cell is exposed to the sun, the circulation of electrons and the appearance of electric current between the two sides of it are produced.

Among the various semiconductor materials used for the manufacture of photovoltaic solar cells, the most used is silicon (monocrystalline, polycrystalline or amorphous), which, doped (artificially contaminated) by a particular element, such as phosphorus, constitutes a semiconductor layer called "n" (with excess negative charge) or a layer called "p" (with excess positive charge) if it is doped by another type of element, such as boron. The union of these two layers (pn junction), provided with the appropriate electrical contacts,

The nominal power of the cells is usually measured in peak watts (Wp), which is the power that the cell can provide with a constant radiation intensity of 1,000 W / m2. For example, the 10 Wp installation would provide a power of 10 W, with a radiation of 1,000 W / m2.

A normal single cell has an area of about 75 cm2 and a nominal power of 2.5 W, which means that, with a radiation of 1,000 W / m2, it provides voltage values of about 0.5 V and current of about 2 TO.

To obtain usable powers by medium power devices, a certain number of cells must be joined in what is called a photovoltaic plate. These plates usually contain between 36 and 72 cells to produce direct current of 12 or 24 V, and provide a power between 80 and 190 Wp.

To optimize the performance of the plates, they must be oriented to the south, with an inclination that depends on the latitude and the time of the year.

Solar energy is composed of three types of solar energy and can be grouped as follows:

PHOTOVOLTAIC

The photovoltaic panel or also known as the photovoltaic solar collector is what we know for the generation of electricity and that thanks to the decree law of the end of the year 2018, which abolished the so-called "sun tax" becomes fashionable again.

The photovoltaic panels or modules are formed by a set of photovoltaic cells that produce electricity from the light that falls on them through the photoelectric effect.

Its operation is governed by the following physical principles. Some of the photons, coming from the sun's rays, impact on the first surface of the panel, being absorbed by various semiconductors, such as silicon

The electrons that lodge in the structure of the silicon are hit by the photons, freeing themselves of the atoms to which they were mainly destined. The movement of this electrons is what we know as electric current, which is generated in a "continuous" way, also called DC, and that we must transform to "alternate" in order to use it in our homes.

THERMAL SOLAR

The solar thermal panel or solar collector is a device that captures the energy of solar radiation for use in heating or domestic hot water. Its operation is very simple, and consists of passing a liquid with antifreeze properties inside.

In its journey through the interior of the collector, this liquid, also called glycol, increases its temperature thanks to the incidence of the sun's rays and the configuration of the panels themselves, which enhance heat accumulation.

Once outside the collector, the glycol will transfer that heat to the sanitary water or for heating, by means of individual exchangers, or inside water accumulators.

PASSIVE SOLAR

Passive solar energy is a set of constructive techniques that enhances the direct use of solar energy through the construction of the building itself. These will allow the transformation of the heat obtained without having to resort to other devices, such as boilers or heaters.

CHARACTERISTICS OF SOLAR POWER SYSTEM

- Source of energy developed and unlimited.
- Energy used to generate electricity and is captured by means of solar panels.
- Energy that is clean by not generating pollution.
- Promote sustainable development
- It is totally ecological and sustainable.

USES OF SOLAR ENERGY

SOLAR POWERED DRIVES

An innovative practice to make effective and convenient use of solar energy is transport that is powered by photovoltaic energy. The most common examples include buses, railroads, cars and even today roads can be powered by sunlight. In fact, solar powered cars are playing a key role in racing competitions around the world, which also speaks of a lot of momentum.

WEARABLE TECHNOLOGY

Consumer electronics is everywhere and is more popular today than ever. It is known to be the most available uses of solar energy. For example, solar chargers have the ability to charge everything from a Smartphone to a Tablet or electronic reader.

There are even solar lanterns that can be charged simply by being exposed to sunlight. Cell phones, wearable, speakers, Tablets, thermostats, as well as dryers or radios, are other examples of the use of solar energy.

SOLAR LIGHTING

It is probably one of the most common uses of solar energy and is also one of the easiest ways to improve energy efficiency in homes. Unlike what happens with traditional outdoor lights, solar lighting does not really require any configuration, since the lights are wireless and take advantage of the sun's energy during the day to avoid the need to provide energy during the night.

SOLAR LIGHTING

Not only that, the aesthetics of solar lighting today can even significantly improve the decoration on the outside of homes. Both the availability and the low cost of these lighting products are one reason why it is so common to find public lighting powered by solar energy.

HEATING

Solar heaters take advantage of the sun's energy and transform it into thermal energy through the use of liquid or air, while solar water heaters use water as a method for thermal transfer. In addition, these solar heating systems can be passive or active. In the first case they are systems that use natural circulation, while with active systems, pumps are used to circulate water and generate heat.

ROOFTOP SOLAR PANELS

Solar energy can also be used to save homeowners a large amount of money by reducing electricity consumption through the use of solar panels. In the end it is important to know that if you want to choose a way of life in which solar energy is used, it is first necessary to estimate the potential savings of solar energy and check the benefits it would provide for the home economy.

The use of energy from the sun to improve the efficiency of housing is advisable, but probably the best way to improve electricity costs is by installing domestic solar panels that have the ability to provide energy to the home without the excessive costs that traditionally have.

PURCHASING A SOLAR ENERGY SYSTEM

If you desire to invest and buy solar panels, the selling prices differ based on the capacity and the size of load.

If you are looking for a loan , ask the following:

- How much will you pay in advance?
- What is the percentage rate that will be applied annually?
- How are your payments calculated?
- Will the amount of fees change over the duration of the loan?
- Will you have to make a lump sum payment?
- Can the lender establish a pledge in guarantee on his house or system?

INCENTIVES AND BENEFITS

If you buy a system, you may qualify for credit for federal, state or local taxes or other incentives. The credit for the federal renewable energy tax for homeowners is equal to 30% of the cost of the system.

This credit will expire at the end of 2016. The Department of Energy has information on specific state incentives for the use of renewable energy.

You may also receive other benefits for the installation of a solar energy system. Depending on the local net measurement rules, your power supply company may pay you the amount of energy your system returns to the grid.

You may also be able to sell the extra electricity produced by your system or obtain credit for renewable energy certificates (REC). A renewable energy certificate is independent of the amount of electricity produced; It is a certificate stating that you generated a certain amount of renewable energy.

When a business, including a business operating in a residential home, which has solar panels sells all renewable energy certificates, it loses the right to tell its customers that it is using renewable energy. It is important that you consider it if you operate a business from home and want to claim that you use renewable energy.

ANALYZE THE PROPOSALS

Compare the detailed proposals of several companies. The proposals should contain the specific details of the system, namely:

- The expected performance of the equipment and the size of the panels.

The total cost of installation, including all charges for construction permits or necessary electrical work.

- If the production of a certain amount of energy is guaranteed

What are the guarantees applicable to the equipment (such as panels and power alternators) and to the labor of the installers.

If you own a solar energy system, you have to maintain the panels and equipment - or pay someone to do the maintenance work - unless the seller includes it in the contract.

Maintenance could include the repair or replacement of the power alternator, or the occasional cleaning of the panels in case of low rainfall. Your equipment may be covered by the manufacturer's warranty for an initial period.

THE COMPANY

When you are looking for a company, ask for references to friends, family and neighbors. Check the background of a company at the corresponding state or local consumer protection agencies and before state boards that issue contractor licenses.

Ask if the company you are considering has the licenses, certificates or guarantees required by the authorities of your state, county or city of residence. For example, your state may require the installer to have an electrician's license.

Also do a search on the internet by entering the name of the company and check what you find.

HOW MANY WATTS DO I NEED

Calculate how many plates you need for your consumption

As we did with the battery size you need, we will try to help you decide how much solar panel power you need. First of all, we will take into account a number of considerations:

The performance of a board is approximately 60-70%, therefore, a 100W solar panel will deliver a maximum of 60W.

The installation position of a solar panel on the roof of a van is not optimal. A solar panel, to work at maximum performance, must be placed at an angle of 45°. If you look, the industrial solar panel installations have that inclination.

The place and time of the year where you are is important. Winter is not the same as summer, just as it is not the same to travel in Andalusia that has more hours of sun, than in Asturias, with much less hours of sun.

That said, let's get calculations done. First of all, to know the power of plates that you must install, you have to calculate the consumption that you will have, as we have explained in this entry.

Once calculated, let's see which board we should install to try to keep the battery charge constant.

The current intensity that a plate will be able to supply is determined by the following formula:

Intensity = Power / Voltage * Efficiency

Therefore, if we have a 100W solar panel and an efficiency of 70%, the formula would be as follows:

100W / 12V * 0.7 = 5.83Ah

That is, the solar panel in full sun, can generate at most 5.83 Amps per hour . This is a theoretical calculation, which although approximate to reality, you should look at the maximum amperage granted by a solar panel before buying it (I should indicate it in the technical specifications).

If we assume that our daily consumption is 50 Amps, let's see what solar panel we would need. To do this, we will also assume that in summer, 7 hours of sunshine per day are used in total, while in winter, due to the weather as there are fewer hours of sunshine, only 3 hours per day on average are used.

Well, assuming we have a 100W solar panel; we would get the following daily energy:

Summer: 5.83A / hx 7 hours = 40.81 Amps per day

Winter: 5.83A / h * 3 hours = 17.49 Amps per day

As you can see, in no case do you reach 50 Amps of daily consumption (a realistic consumption). And that, taking into account other losses that exist (such as the loss of the charge regulator), the loss in the wiring or that it may be that for days there is not even a sunbeam ... Therefore, with 100W, The solar panel will simply be a support for the battery to discharge more slowly.

Another thing would be to install solar panels of 200W or more capacity (two of 100W can also be installed). With these figures, one could start talking about energy self-sufficiency through solar energy in a camper van. (With 200W, in summer 80A would be generated daily, and in winter about 35A daily).

COMMON MISTAKES WHEN INSTALLING SOLAR PANELS

Solar panel is an invention that is helping the planet, and our pockets. Investing in a system of solar photovoltaic panels at home contributes to maintaining a great saving in the receipt of CFE.

The interest in the subject, of clean energy, is growing every day, and doubts always arise, questions when looking for an installer or supplier of solar panels, that is why it is important that you take into account these common mistakes when buying panels solar so you don't make them when you're in that process.

SOLAR PANELS HAVE VERY HIGH PRICES

People think that the panels are highly expensive, however, they have to be seen as an investment, not as an expense. There was a time when the panels were too expensive and it is a date that people still see it as a luxury purchase, but they must consider that having solar panels at home contributes to generating an impressive saving on the CFE light bill.

It has always been considered a serious investment for companies and institutions that were looking for savings, even residential buyers preferred not to know the subject because they thought it was quite expensive. But the cost of solar panel systems has been declining in recent years, and in the same way their performance and efficiency has improved in an indescribable way.

Apart from that the government and other institutions have shown funding support to promote the use of solar panels both in companies and in homes. Thanks to this there are already many companies and homes that save money by taking care of the environment by having a system of photovoltaic solar panels.

When you go to make a purchase of solar panels, the first thing your installer has to do is generate a quote, this is based on your current electricity bill, and the panels that would be installed, from here they can provide you with the performance and savings What would you have if you installed the solar panels.

ALL SOLAR PANELS ARE THE SAME

There are thousands of brands and types of solar panels around the world, and it is quite difficult to tell the difference between various types of solar panels. Typically, companies and companies offer the same information about solar panels; starting with the 10 year product warranty, then the 25 year performance guarantee and a nominal power, which is usually around 250 WP.

Therefore the best way to say if a solar panel is good or not, is to ask for references in other countries, and check if there are tests on those solar panels. If there is evidence that certain solar panels are used in large projects in various markets, it means that the solar panels will work perfectly.

THE COUNTRY OF ORIGIN AFFECTS THE QUALITY OF SOLAR PANELS

There is a common ideology between companies and consumers that German-engineered solar panels are better than others. However, judging a product by its country of origin may become only a strategy to convince the consumer through ideologies already raised with other products, and emotionally motivate them to achieve a purchase of solar panels.

The photovoltaic industry grew exponentially in recent years, therefore some manufacturers could barely adapt, others had to look beyond and had to open facilities in other countries, all this to grow solar panel production.

Therefore, it is not fair to assume that the quality of a solar panel has something to do with the country from which it is being manufactured. For example, It is believed that a solar panel manufactured in China is less efficient than one manufactured in a non-Asian market.

However, more than 90% of solar panels imported into Australia in 2013 were made in China. With such an impressive market penetration level, it is hard to believe that Chinese solar panels are of inferior quality.

TECHNOLOGY IS UNSTABLE

Another common misconception about solar panels is that the technology is fairly new, and it is better to wait a few generations before buying them. What most people don't know is that the technology behind photovoltaic modules has been around for more than a century.

While it is fair to say that unstable photovoltaic systems for a long time, recent events have increased the performance of solar panels. And, with the growing interest in clean energy, there is no better time to invest in solar energy than now.

SOLAR PANELS ONLY WORK IF THERE IS ENOUGH SUN

Most people are wrong about the fact that solar panel only work in direct sunlight. If your business is in an area where there is not much sunlight, then it is not worth investing in electricity.

But, here is the truth. While it is true that most solar modules are more efficient in the sunnier weather, solar panels can still generate a considerable amount of energy even in cloudy conditions.

You have to keep in mind that most photovoltaic systems are installed in addition to an existing power supply in the network. Therefore, a rainy day will not mean that your company will remain without electricity.

Just look at Germany, a country with a fairly rainy climate - they are a solar superpower, getting most of its power from photovoltaic systems. Apart from that the installers must analyze the area where they are going to be accommodated, and they must give the best positioning to the panels so that they can generate a good amount of energy.

IT'S HARD TO FIND A RELIABLE INSTALLER

With such an overwhelming list of manufacturers and installers, it can be difficult to say the good of the less professional. Apart from the generic information, such as "we have installed 5,000 systems", it is difficult for consumers to assess the quality of the service experience they are buying.

As a general rule, when you decide on your solar panel supplier, ask for references, preferably from the companies that have already attended.

The decision to install a photovoltaic solar system is one of the best that will bring numerous benefits. But, before buying photovoltaic modules, research and find an installer that can meet your specific needs.

CHAPTER THREE

PHOTOVOLTAIC SYSTEM

A photovoltaic system, in simple terms, is the grouping and joint work of certain electrical components to achieve the transformation of solar energy into electrical energy usable for any conventional electrical device or device of a house, a business or even an industry.

To achieve the generation of this type of renewable energy, the basic electrical components of a photovoltaic system are: solar panels, inverter, charging center and bidirectional meter.

Solar panels or photovoltaic modules are the ones that usually take the applause and glory in the task of generating solar energy, however, if we thought of solar panels as players of a soccer team, they are only part of the team.

It is the joint work of all the elements that make it possible for the team to function properly for everyone to fulfill their assigned function.

The generation of electrical energy will depend on the hours that the sun shines on the solar panel and the type and quantity of modules installed, orientation, inclination, solar radiation that reaches them, quality of the installation and the nominal power.

The devices through which solar energy is absorbed are solar cells. These are elements of photovoltaic systems that have the ability to produce electricity by harnessing the sunlight that affects them. Once the photons emitted by the solar radiation come into contact with the atoms present in the solar cells, electrons are released that begin to circulate through the semiconductor material with which the cells are manufactured and electrical energy is produced.

A photovoltaic system can be "interconnected" which is the most convenient for residences or businesses with access to the CFE's power grid. With this system the generated energy is injected into the electricity grid and from there it is taken when one needs it.

The other option is an "island" system that allows the supply of electricity in places inaccessible to the electricity grid. These systems are mainly used in country houses or in telecommunication antennas.

When the photovoltaic cell is reached by the light, it releases electrons, thus creating an empty space called an egg, that hole will tend to be filled by another free electron that was also excited by the radiation of the light that reaches the semiconductor, in lines. In general, the idea is that the holes move for one side and the free electrons for the other, in this way a current flow will be generated on the semiconductor.

MAIN COMPONENTS OF PHOTOVOLTAIC SYSTEMS

A photovoltaic solar energy system, also called a solar energy system or a photovoltaic system, is a system capable of generating electrical energy through solar radiation.

A photovoltaic system has four basic components:

Solar Panels - They play the role of heart, "pumping" energy into the system. They can be one or more panels and are sized according to the energy required. They are responsible for turning solar energy into electricity.

Load controllers - They act as valves for the system. They serve to prevent overcharging or overcharging of the battery, extending its life and performance.

Inverters - The system's brain is responsible for transforming the 12 V of direct current (DC) of the batteries into 110 or 220 V of alternating current (AC), or other desired voltage. In the case of connected systems, they are also responsible for synchronization with the power grid.

Batteries - They work like lungs. They store electricity so that the system can be used when there is no sun.

While an isolated system requires batteries and charge controllers, grid-connected systems work only with panels and inverters as they do not need to store power.

In general, photovoltaic systems usually have the following elements:

- - Solar cell modules
- - Structure for the modules
- - Operating instruments
- - Voltage regulators and controllers
- - Electric storage batteries
- - Switches and wiring
- - Power grid around

OPERATION OF A PHOTOVOLTAIC SYSTEM

The operation of a photovoltaic system is possible thanks to the solar panels where, thanks to the photoelectric effect, solar energy is converted into direct current electrical energy, which cannot be used conventionally if it is not transformed into alternating current.

This is where the function of the inverter, a key part of the photovoltaic system, comes into play, since it is he who converts the current to be compatible with any type of installation, such as domestic, commercial or industrial.

Subsequently, depending on the type of photovoltaic system, you can have charge controllers that regulate the use of energy and a battery bank that allows energy storage.

HOW DOES THE PHOTOVOLTAIC SYSTEM WORK

Do you know how the photovoltaic system works? The photovoltaic system is based on the use of photosensitive panels that are able to transform the energy of the sun's rays into electric current (direct current).

The generated power is sent to the inverter which is the equipment responsible for converting the energy to the utility grid standards (alternating current). Subsequently, the energy is injected into the home's power grid and can be used by the consumer.

All surplus energy not immediately used by household equipment will be transferred to the utility network. This energy coming from the installation site passes through the energy meter clock. This portion of energy becomes credits.

Thus, the user can use in periods when the system is not generating power, at night, for example. It can also be used when power consumption is higher than generation. In the latter case, the solar system provides part of the energy and the utility supplies the rest of the energy being consumed.

TYPES OF PHOTOVOLTAIC SYSTEMS

There are two basic types of photovoltaic systems: Off-Grid Systems and Grid-Tie Systems.

Isolated Systems: are used in remote locations or where the cost of connecting to the power grid is high. They are used in country houses, refuges, lighting, telecommunications, water pumping, etc.

Grid Connected Systems: on the other hand, replace or complement the conventional power available in the grid.

Depending on human needs, the photovoltaic system may vary in type, capacity or specifications.

CONSIDERATIONS FOR INSTALLING A PHOTOVOLTAIC SYSTEM

Only through the proper functioning of the photovoltaic system in which a specialized team of engineers and technicians with knowledge and experience in all the devices involved participates, can energy from the sun be generated safely, conveniently and efficiently; sustainable and environmentally friendly.

To obtain the necessary components for a photovoltaic system, professional solar energy installers can turn to specialist wholesalers such as SDE Mexico who, in addition to providing these components, can guide and guide them in new technologies and correct use of them to achieve a successful installation of the system.

ADVANTAGES OF PHOTOVOLTAIC SOLAR ENERGY

It is inexhaustible: We can consider the sun as an inexhaustible source of energy, its rays will reach the earth while the planet exists, in this way it is logical to consider it as an inexhaustible source of energy.

It is Clean: It does not emit any type of pollutant to the environment

Ideal for remote areas: It is the appropriate technology to supply electricity to areas where the power line does not reach or is inaccessible, for example remote rural areas, islands or small cities

It is everywhere: In any part of the world where the sun shines we can have access to this technology, it is a very important advantage since it gives us independence from the important implementation area, if we compare it for example with the hydroelectric dams that They can only be installed on highly flowing rivers, it represents a great advantage.

More advantages include:

- As it comes from a renewable energy source, its resources are unlimited.
- Its production does not produce any emission, that is, it is a very environmentally friendly energy.
- The operating costs are very low.
- Maintenance is simple and low cost.
- The modules have a life span of up to 20 years.
- Not only can it be integrated into new construction structures, but also into existing ones.
- Modules of all sizes can be made.
- The transportation of all the material is practical (with this reference is made that unlike for example wind energy, where the transport of the

material is complex due to the size, the material used in photovoltaic energy is easier to transport).
- The cost decreases as technology advances.
- It is an ideal energy use system for areas where electricity does not reach.
- The photovoltaic panels are clean and quiet, so they can be installed almost anywhere without causing any discomfort.

DISADVANTAGES OF PHOTOVOLTAIC SOLAR ENERGY

LARGE INITIAL INVESTMENT

The costs of the initial investment are high, although over time they are amortized, a large amount of money is needed to face the first stage of investment, perhaps for a small household with low demand the cost is more reduced but also represents a high value.

LARGE TERRITORY DESTINED TO PLACEMENT OF PANELS

Like wind energy, if we want to implement a system for large consumption, at the level of a small city for example, we need a large area of land destined for the placement of solar panels, and this can be a problem if you don't have that space.

INSTABILITY OF SOLAR RADIATION

Depending on the area, the time of the year and the weather the amount of radiation can only vary, thus making the amount of solar energy that we can store unstable, this can be a problem if we do not have sufficient storage capacity (batteries) to cover the low solar radiation season.

CHAPTER FOUR

PASSIVE SOLAR SYSTEM

Passive solar systems are those that allow the collection and accumulation of heat that comes from solar energy through elements such as windows, walls or roofs. To do this, they will not need to use other electromechanical devices such as pumps or fans, which explain their denomination of liabilities.

Passive solar designs are based on the search for the entry of light, as well as good ventilation and insulation, making the sunlight provides heat in winter:

A well-designed system will maintain the temperature in the summer months, without overheating, thanks to the air currents and the thermal mass level of the material chosen for the walls.

Thermal mass is an essential element within this scheme. Concrete, stone, tiles or brick are some of the materials that offer better results for passive solar homes, given their ability to absorb and retain heat not only in the hot seasons but in the colder

To the above it is necessary to add windows or other surfaces that allow a use of solar lighting both for the capture of energy and so that there are no overloads

Good interior insulation and mechanisms that transfer heat, as well as allowing it to control, as far as possible, will cause the temperature of a home to stabilize, without expenses and without environmental impacts

It is an alternative, which is increasingly extended and that, in combination with active solar energy in office buildings, requires an initial investment to build or adapt the building.

However, it should not be forgotten that the initial expense is compensated by the low maintenance cost of this system as well as by the savings it allows for invoices. That is why it is necessary to assess an alternative that allows you to go a step further by investing in a more efficient energy model.

HOW PASSIVE SOLAR IS USED

In this way the capture and accumulation of energy from the Sun is achieved in a discreet and economical way, since they are systems belonging to the same design of the building. Moreover, it is a resource widely used in bioclimatic architecture

Its ability to insulate indoor outdoor environments is an aid to avoid strong temperature contrasts. Basically, this is possible thanks to the accumulation of heat, to subsequently transfer it to the outside environment, precisely when temperatures fall outside

Passive-type Sun-generated energy uses components of the construction style of walls, windows, ceilings, floors, or exterior building elements and landscaping to control the heat generated by the Sun:

The goal of solar heating designs is to capture and store the thermal energy of direct sunlight

A relevant aspect regarding the design of buildings destined to take advantage of passive solar energy is based on knowing how to manage the flow of air currents inside the building in order to capture heat in winter, dissipating it outside in summer

The main solar concepts are natural lighting, passive heating and cooling

An optimal architectural design also allows the improvement of natural lighting. An excellent way to lower the electricity bill of a commercial building is based on the use of natural light.

In addition, it will be possible to create a pleasant environment while reducing the cost of air conditioning, since the amount of heat generated by light bulbs and artificial power supplies is considerable.

DISADVANTAGES OF PASSIVE SOLAR SYSTEM

Reliability is one of the main drawbacks of passive solar is that house heating is completely dependent on the weather. Many passive solar designs include the ability to minimize the warmth of the sun and maximize heating, but bad weather can make the home cold and boring

It is essential to evaluate the suitability in each case and the optimal design combined with a different style of traditional or renewable energy source. Not surprisingly, the goal is to achieve the lowest possible energy costs through production that meets needs and is considered environmentally friendly.

HOW TO USE PASSIVE SOLAR

The energy efficiency of passive solar systems is based on the basic mechanism of heat transfer, namely the use of natural resources based on convection, conduction and radiation.

In this way, the collection and storage of energy from the sun is realized in a cautious and economical way because it is a system belonging to the same design of the building.

In addition, it is a widely used resource in bioclimatic architecture

The ability to insulate indoor outdoor environments helps to avoid strong temperature contrasts. Basically, this is possible thanks to the accumulation of heat, and then accurately transfers it to the external environment when the temperature goes out.

Passive solar energy controls the heat generated by the sun using walls, windows, ceilings, floors, or external building elements and landscaping structural style components. The goal of solar heating design is to collect and store direct sunlight thermal energy.

The relevant aspects of designing buildings that will use passive solar energy are based on knowing how to manage the flow of air in the building to take heat away in the winter and dissipate outside in the summer.

The main concept of the sun is natural light, passive heating and cooling

Optimal architectural design also improves natural light. An excellent way to reduce electricity bills in commercial buildings is based on the use of natural light.

In addition, the amount of heat generated from light bulbs and artificial power sources is quite large, creating a comfortable environment while reducing the cost of air conditioning.

DIFFERENCE BETWEEN ACTIVE AND PASSIVE SOLAR ENERGY

There are basically two ways to use energy from the sun:

Find out more about solar thermal energy through active solar (solar power and solar thermal energy).

Through passive solar

The main difference between active solar energy and passive solar energy is that active solar energy has a process of energy conversion.

Through solar panels, solar energy is converted into electrical energy, and through solar panels, solar energy is converted into thermal energy.

The energy generated from the passive sun is called a technology that can be used directly without processing solar energy.

PASSIVE SOLAR ENERGY FOR CHILDREN

The recognition of the new generation from the younger generation about the importance of caring for the earth is a challenge that every adult must face. Knowing more aspects about passive solar energy will definitely help you explore interesting systems that will help you maintain a close and remote environment.

APPLICATIONS OF PASSIVE SOLAR POWER SYSTEM

Passive solar technology includes systems with direct and indirect gain for space heating, thermo-siphon, based water heating systems, the use of thermal mass and phase change materials to soften air temperature oscillations, solar cookers, solar chimneys to improve natural ventilation and the earth's own shelter.

It also includes other technologies such as solar ovens or solar forges, although these require some energy consumption to align concentrating or receiving mirrors and have historically not proven to be very practical or cost effective for extensive use. This is relative to photovoltaic energy.

CHAPTER FIVE

CONSERVATION RULES

The biggest challenge of construction today is energy efficiency. The need to reduce energy consumption in our homes and work environments is not something new, but has been one of the trends in the sector for some time, but now more than ever it is essential to bet on it. The European Union has forced Spain to catch up and, as of 2020; all the buildings that are built will have to be almost zero consumption building.

However, beyond the legislation, it is essential that users and professionals become aware of the importance of energy efficiency in buildings , since beyond the myth of costly investment in adapting to a sustainable model in the consumption of supplies, We can save a lot on electricity or gas bills throughout the life of the building.

In addition to contributing to an improvement of our planet, these are some of the benefits of energy efficiency:

Rehabilitating a home to be energy efficient can save 40% on the electricity bill.

The habitability conditions of the buildings are much better in conditions of energy efficiency. For example, there are no excesses such as overheated stays in winter or very cold in summer.

Normally, noise is also much lower when living or working in a space that is sustainable. Without the use of air conditioners the noise is reduced, as well as being able to have the windows closed and much more watertight outside.

If everyone opted for sustainable energy models, cities would certainly be much cleaner. It is not cars or industries that most dirty the atmosphere, but buildings are the main focus of pollution.

In order to have energy efficient houses we need to take into account several factors, equally. It is not enough to launch a single proposal, but we must know that it will be the combination of the different systems that will maximize the results.

For example, proper air conditioning through the use of a mechanical ventilation system is necessary in energy efficient homes, but for its potential to be maximized it is also necessary that each room is sufficiently tight.

Another very important factor is the energy consumption of household appliances, lighting, water heaters or kitchen stoves. All these systems are easy to change and it is very important that, whenever we do with new appliances at home or appliances, we must keep in mind that not only you have to look at its design or its price, but also its consumption.

An energy saving light bulb can cost up to twice the cheapest of the traditional ones, but its useful life can be up to three times longer. For that reason alone, the price difference compensates. But it will also spend a small part of what the traditional needs.

ENERGY EFFICIENCY AND BUILDING CONSTRUCTION

The first thing to keep in mind to achieve buildings that meet the minimum energy efficiency is its construction. In the middle of the real estate bubble, at the beginning of the century, the use of air conditioning and heating equipment became popular.

It was at that time that many houses were built with walls that allowed cold and heat to pass through, poorly insulated windows and materials that did not guarantee a good maintenance of home comfort. Heating and air conditioning systems were installed to give a feeling of thermal comfort that, in fact, meant a very high supply bill.

Today, this cannot be done. In order to bet on energy efficiency, the way in which our buildings are constructed is paramount. You have to bet to maximize the tightness of all the rooms that are built.

To improve this tightness, the following can be taken into account:

The walls must be constructed of resistant materials and do not allow heat to escape in winter or to get hot through them in summer.

The windows and doors must be tight and very resistant. It is useless to have a large balcony if the window allows all the cold to pass in the hardest months of the year.

In the summer months, it is possible to implant awnings, ventilated roofs or ventilated facades. With this we will give an air passage that will prevent the stay from overheating. In winter, its effect also helps.

If a home is built in the most appropriate way, taking all this into account, we may not have to use air conditioners for most of the year.

ENERGY EFFICIENCY AND VENTILATION SYSTEMS

Ventilation systems with constant flow double flow

In order to minimize the use of air conditioning systems, double flow mechanical ventilation is not only necessary, but also helps us greatly improve our health.

Mechanical ventilation systems will be necessary in the almost zero consumption building. However, those that are optimal for guaranteeing the highest degree of energy efficiency are double flow. In this case, the windows must be closed at all times and ventilation is performed automatically.

Dual-flow mechanical ventilation eliminates stale air or whose quality has worsened, and introduces air from the outside, previously filtered. Thanks to this, health at home is much better thanks to breathing an air of the highest quality. For that reason alone, these systems are necessary in modern building.

Even so, in energy efficiency they have great advantages. The first is that, by introducing outside air, double flow ventilation has a bypass system that heats the air that is introduced from the outside in winter thanks to the passage through the ducts.

That means that, without any cost, hot air is already being introduced in winter. No need to use heating. That, coupled with a good tightness of the building, eliminates our need to turn on the expensive air conditioning systems.

In summer the system is also able to improve thermal comfort. In that case, the air introduced is cooled.

Thanks to this, mechanical ventilation is able to offer huge improvements in energy efficiency.

ENERGY EFFICIENCY AND LOWER CONSUMPTION APPLIANCES

The first steps to take with energy efficiency came with the light bulbs. Nowadays, it is already very difficult to see traditional bulbs of great consumption and they are only common to renovate old installations or traditional lamps that, by size, can only use those.

Likewise, it is recommended that in our homes we change the entire facility that uses high energy consumption for another that turns out to be sustainable. For example, light bulbs are a clear example. But, depending on how, you can also install ceiling fans for situations where we want to cut electricity consumption and the use of air conditioning.

When choosing appliances, televisions or any device that we use at home, we should look for it to be low consumption. However, what about hot water and hotplates at home?

Sanitary water, today, can be heated in various ways depending on energy efficiency. At a minimum, it is recommended to use efficient heating boilers. And in the case of household stoves, which are electric and low consumption.

RECOMMENDATIONS TO IMPROVE OUR ENERGY EFFICIENCY

Until a few years ago, the requirements of the energy label were minimal and many builders were limited to comply with the requirements of the Law, without any improvement beyond the minimum. Today, it can't be like that. In addition, for the end user it is worth paying a little more and being able to have sustained savings throughout the life of the building.

In any case, in our homes we can begin to take into account many aspects to implement a strategy that, in a simple way, begins to improve the energy efficiency of our homes. We recommend the following:

Try to find the air currents. If you see that in a window, in a joint, in a door or even on a wall, the wind sneaks into the house, you already know that you have a leak. Ideally, cover that joint or improve the tightness in that area.

Do you use gas to heat the water? Consider switching to accumulation boilers. They are much more efficient. Electricity can also be a saving when cooking.

Protect your windows and doors if they are many years old. With adhesive sheets, the specialist can cover the joints so that the heat stays or not in the cold.

Of course, the sink is a primary place. The taps must be single lever and the double push-button basin.

In summer, use awnings to stop the heat and try to ventilate at night, when the temperature is lower. The use of air conditioning should never be below 24ºC. It will not cool faster, and it will be a great unnecessary expense.

If we bet on energy efficiency in homes, the biggest beneficiaries will be the users. But also cities and the environment will achieve a great improvement.

CHAPTER SIX

OFF-GRID SOLAR PV SYSTEM

Off-grid solar systems use solar panels to charge a battery or a group of batteries, then you use the energy stored in batteries to power your lights and appliances.

The solar systems with network connection are the ones you find in the homes and businesses of your city. They interact with the power grid and do not require the use of batteries.

The type of off-grid solar system that we will analyze here is the type of kit you can use to go camping or the type of systems you find in recreational vehicles and ships.

SOLAR PANELS: YOUR POWER SOURCE

Typical single 12V panels: when you are charging individual batteries or a small group of them, it is recommended to use low voltage solar panels. These panels vary in size from 3W to 130W and have a power of 12 or 24 volts.

A 60W panel can supply enough power to have a pair of efficient lights on at night and charge some electronic devices such as cameras and cell phones during the day.

If you plan to charge laptops or use a small DC refrigerator, I suggest you start with an 85W panel. If one panel is not enough, you can add another panel,

COMPONENTS OF AN OFF-GRID SYSTEM: SOLAR PANEL

12V / 10A charge controllers: the charge controller is the most economical component of your system and will protect your expensive batteries. Charge controller will protect your batteries from permanent damage.

Selecting the correct charge controller is quite easy too. You must first select the voltage of your system (12, 24 or 48 volts). Some load controllers include a field configurable voltage function that will allow you to select and change the operating voltage.

DEEP CYCLE BATTERIES: YOUR ENERGY STORAGE

12V deep cycle batteries: batteries and especially battery maintenance are very extensive issues. For a simple solar system like the one we are trying to put together in this article, you should use commercial 12V deep cycle batteries or marine grade batteries.

Do not use car batteries. Shipping batteries is expensive, so try to buy them at local locations. You can even search in places like Costco, Walmart or a local renewable energy provider. These batteries come in 12 volts.

When you have a 12V system and want to increase the energy storage capacity, you can connect several batteries (no more than 4) in parallel. When your system is operating at 24 or 48 volts, you must connect pairs of batteries in series to reach the desired voltage.

If you want to increase the quality and durability of your battery bank, you might consider using industrial grade industrial cycle batteries.

TYPES OF SOLAR PANELS

MONOCRYSTALLINE PANELS

The mono-crystalline solar panels are those with older technology, but is the most developed to date. They are made of a single crystal of pure silicon, as the name implies. These types of panels have the highest efficiency rates, since they are manufactured with high purity silicon. The efficiency in these panels is above

15% and in some cases exceeds 21%. In addition, they usually work better than polycrystalline panels of similar characteristics in low light conditions.

They have a long shelf life; most manufacturers offer a 25-year warranty on their mono-crystalline solar panels. But all this makes them a little more expensive than the rest.

POLYCRYSTALLINE PANELS

The manufacturing process of poly-crystalline photovoltaic panels is simpler, which results in a lower price. Instead of going through an expensive and slow process, manufacturers simply put a crystal seed in a molten silicon mold and allow it to cool; this is the reason why the crystal surrounding the seed is not uniform.

Assessing the economic aspect, for domestic use it is more advantageous to use polycrystalline or even thin layer panels, although their efficiency is lower compared to mono-crystalline ones (round of 13 to 16%). For that reason, it would be necessary to have more space, to be able to accommodate larger panels and thus achieve the same performance as with mono-crystals.

THIN FILM PHOTOVOLTAIC SOLAR PANELS

These panels are lightweight and generally immune to shading problems or obstructions and low light conditions, which do not usually hamper their performance. They can be flexible, allowing them to adapt to multiple surfaces.

COMPOSITION OF A PHOTOVOLTAIC PANEL

The solar panels are sets of photovoltaic cells each producing approximately 0.5 V. connected in series and in parallel, the photovoltaic cells can produce a voltage of several tens of volts with a current of a few amperes .

In these cells a metal grid is placed, which in turn, are attached to a conductive plate. These two elements allow to collect the electric current created by the cells during their exposure to light.

A transparent plate , often in tempered glass , covers the panel without diminishing the light contribution. This plate is protected from shocks with a frame that allows a solid anchor to a roof or a structure on the floor. Even with a lot of wind, nothing should move to ensure safety.

COMPONENTS OF A SOLAR ENERGY SYSTEM

The solar panel is the main component of all types of photovoltaic systems. Depending on the application, there are various parts that add to the system.

SOLAR MODULE (SOLAR PANEL) PHOTOVOLTAIC

Component responsible for transforming solar radiation into electrical energy through the photoelectric effect, they are made mainly by semiconductors (silicon) mono-crystalline or poly-crystalline.

The ones with the best price and the highest availability in the international and Colombian market are the polycrystalline. These are characterized by their nominal power or maximum power that this panel can generate under ideal conditions (radiation of 1kW / m2 and temperature of 25°C).

CHARGE REGULATOR

This component of the system efficiently manages the energy to the batteries, prolonging its useful life, protecting the system from overload and over-discharges. This component is commercialized based on its maximum current capacity to be controlled (amps).

BATTERY (ACCUMULATOR)

The electrical energy of the panels, once regulated goes to the batteries. These store electricity to be able to use it at another time, its commercialization is based on the ability to store energy and is measured in Ampere hour (Ah).

INVESTOR

This component converts the direct current and low voltage (12v or 24v typically) from the batteries or controller into alternating current, in the case of Colombia 120 V, in a simplified way it can be said that it transforms the direct current into a conventional outlet. It is usually marketed based on its power in Watts, which is calculated as the voltage per current (P = VI).

It corresponds to the maximum demand of (power) of the equipment to be connected. This component can be dispensed with when the equipment to be connected can be powered by direct current. As is the case with some types of lighting, motors and equipment designed to work with solar energy.

BRACKETS

This is a passive component of solar energy systems. Responsible for keeping photovoltaic modules in place and must be designed to withstand constant weather, thermal expansions for at least 25 years.

Each of the previous components of a solar energy system uses different technologies. Which make the systems more or less robust and provide other types of properties?

The use of each of these components and the technology to use depends a lot on the need. What is sought to cover and technical limitations? In other words, if you want a portable system, you should reduce the weight of the batteries, the most convenient thing is to use lithium-ion batteries. In cases of very high humidity, encapsulated controllers with a high degree of water protection must be used.

FIRST AND FOREMOST: CHOOSING AMONG PHOTOVOLTAIC MODULE TECHNOLOGIES

To transform the sun's energy into energy that we can apply to our daily life, we will need a photovoltaic cell or cell.

Solar or photovoltaic cells are small cells made of semiconductor materials, such as crystalline silicon or gallium arsenide, which can behave as conductors of electricity or as insulators, depending on the state they are in. Generally, the solar panels that you will find in the market are made with silicon.

A solar cell is not capable of generating large amounts of energy alone, so several of them are combined and a solar panel is formed. They can be 36 cells or more, depending on the size and power required of the photovoltaic solar panel. Therefore, a solar panel is actually a large plate in which there are many solar cells together. If a cell converts the sun's energy into electricity, a panel allows generating enough energy to use in a house.

HOW DO SOLAR PANELS WORK

In order to catch the energy coming from the sun and convert it into electrical energy, there has to be a process in which several parts intervene. First, semiconductor crystals receive a treatment that seeks to give each one a positive charge and a negative charge.

In this way, it is achieved that the cells have both charges and can generate electricity. Then, they are placed on the panel intercalating them and linking them together, using a conductive thread.

When these crystalline cells are directly exposed to light, the sun's energy causes the electrons in the negatively charged part of the cell to move towards the positively charged part. In this way, thanks to sunlight and the materials used to assemble the cells, we generate an electric current from one point to another. All together they produce an electric field in the solar panel.

HOW TO USE SOLAR ENERGY AT HOME

Once the solar energy is captured and transformed by the cells into electrical energy, that current must be transferred and adapted to the household demands. You can read: Uses of solar energy that you can take advantage of today .

Beyond solar panels, there are a number of basic components that complete a photovoltaic system:

INVESTOR

It is the heart of the system, where electricity is managed based on demand and production. This device transforms the direct current of the accumulator into alternating current at 230V 50Hz. Deliver the necessary energy at all times. Ask for help from external sources, due to excessive demand or protection of the accumulator, managing the battery charge and operating in the latter case as a charger.

BATTERIES

The electrical energy of the panels, once regulated, goes to the batteries, which are the ones that store the electricity to be able to use it at another time. Marketing is based on the ability to store energy and is measured in Amps hour (Ah).

CHARGE REGULATOR

This element has the function to protect the battery in case of overcharging or deep discharges which affects the storage system minimizing its useful life. The regulator constantly monitors the battery bank voltage when the battery is charged interrupts the charging process by opening the circuit between the panels and the batteries, when the system begins to be used and the batteries to be discharged the regulator reconnects the system.

HOW A SOLAR PANEL PRODUCES POWER

These devices are made of two types of semiconductor materials, one of positive charge (p) and one of negative charge (n). When exposed to light, they allow a photon of sunlight to "start" an electron; the free electron leaves a "hole" that will be filled by another electron that in turn was torn from its own atom.

The job of the cell is to cause free electrons to go from one semiconductor material to another in search of a "hole" to fill. This produces a potential difference and therefore an electric current, that is to say, a flow of electricity will be produced from the point of greatest potential to that of the least potential until at two points the potential is the same.

This phenomenon, the photovoltaic effect, manifests itself in the form of a current of a few milli-amps .

THE 3 TYPES OF PHOTOVOLTAIC PANELS

Many photovoltaic cell technologies are distinguished by their price and performance. The higher the latter, the greater the maximum power and therefore the annual output. The most efficient technologies are intended for specific scientific applications.

AMORPHOUS SILICON

A square meter (1 m^2) of these panels produces approximately 60 Wp, a yield of approximately 6% . In addition to their low price, they work even in diffused light and can be found in flexible supports , more practical for nomadic use (without fixing it to a surface).

POLYCRYSTALLINE SILICON

Its maximum power is around 100 Wp per square meter (m^2) or a yield of approximately 15%. When we look at them, we see blue spots of different shades. Its performance is lower with cloudy weather.

MONOCRYSTALLINE SILICON

More complex to build than the previous ones, they are also 50% more powerful (16 to 24% yield) or approximately 165 Wp / m^2. They are also sensitive to diffuse light that causes their performance to decrease.

HOW TO MEASURE THE POWER OF A PHOTOVOLTAIC PANEL

The power of a solar panel is expressed in peak watts (Wp) . This is the power provided by a solar panel at 25 ° C, with sunlight and a power of 1000 W per square meter.

This standard size makes it possible to compare different panels between them, the ones with the highest peak power produce, of course, more electricity. This allows the annual production of a photovoltaic installation to be evaluated according to the installation site. For example, with a kWp panel, you can expect to produce in a year between 1000 kWh and 1500 kWh in Bilbao in Murcia.

WHAT SIZE SOLAR PANEL TO CHOOSE

The size of a solar panel depends on several parameters that you must take into account. You can choose it depending on your objective: the maximum power per square meter (m^2) of solar panel installed, which is a function of the chosen technology and the intrinsic performance of the chosen model; the amount of energy you want to produce in a year, expressed in kWh of the amount of sun , and therefore, its location.

The area available for the installation of photovoltaic panels;

Your orientation ideally they should face south and form an angle of 35 ° with respect to the horizontal. The more the panels deviate from these optimal conditions; they will produce less electricity with the same amount of sunlight. A panel placed vertically or horizontally will never produce as much as it could with ideal conditions.

CHAPTER SEVEN

ABOUT INSTALLING PANELS AT HOME

Current consumption: Do you know how much you consume Kw / h per day? Did you do everything possible to make your consumption more efficient? Do you use the air at 24°C? Do you unplug your phone / PC chargers when you are using them? Is your home lighting LED technology?

The cheapest / most economical energy is the one that is not consumed! The more you have paid attention to this, the better prepared you will be to take the step towards a solar installation in your home. You can check your electricity bill, or check with the electric company. This will help you make your choice by calculating your consumption.

Orientation: The panels work by maximizing the generation of energy when they receive the greatest amount of solar radiation throughout the day. For the Southern Hemisphere, that means orientation to the north, for the Northern Hemisphere it means an orientation towards the South. However, other addresses are also viable, they will simply have less performance.

Shading: You should also check if your roof receives shade from trees, buildings or other objects. The most important hours where you should receive light are from 10am to 3pm.

ROOF CONDITION BEFORE SOLAR INSTALLATION

The roof is the best place to install the Solar panel systems. To determine if a roof is suitable for solar energy, solar installers will analyze:

ROOF ADDRESS

The south-facing roofs are ideal for solar energy, as they receive the most sunlight during the day. However, a solar panel system on a roof facing east or west can still produce enough energy to reduce your electricity bills, save money and reduce your carbon footprint. Learn more about how roof orientation affects the production of electricity.

TILT ROOF ANGLE

Solar panels should be installed on roofs with an inclination between 15 and 40 degrees. Even if your roof is flat, it can still become solar as long as you mount your panels at a good angle. Learn more about the impact of the roof angle on electricity production.

THE SIZE AND SHAPE OF ITS ROOF

It is easier to install panels on a large square roof. A general rule is that for every kilowatt of panels you want to install, you will need approximately 100 square feet of roof space. A typical domestic system will require approximately 500 square feet of space to install solar panels.

Things like skylights, dormers and fireplaces will affect the amount of space available for a solar installation in the home. In general, solar installers can design the design of their panels around these obstructions so as to maximize their electricity production.

SHADING ON YOUR ROOF

It is important that your roof receives enough sun during the day to maximize electricity production. Tall trees or nearby buildings can block the sun, cast the shadow on the roof.

AGE OF YOUR ROOF

Solar panels are warranted at 25 years, and removing them to replace your roof a few years after installation can be expensive. Before installing your panels, make sure your roof is in good condition and it is not necessary to replace it in the near future.

INSTALL A PHOTOVOLTAIC SOLAR SYSTEM

Mounting solar panels and placing the devices is easy

Installing a small solar system in your home is not difficult and some prefer to do it without help. Not only can you save, you also learn. We do not oppose this idea, on the contrary. But for reasons of safety, quality and guarantees, it is preferable that qualified personnel accompany them, avoiding any problem.

PREPARATION OF A SOLAR PANEL

If you dare to do it alone, all responsibility, including legal and risk is yours. It can affect the guarantees. We are not responsible for any damage that may occur.

All quality equipment comes with an installation manual where procedures are indicated and how to adjust them. Sometimes, these manuals exist only in English. You have to follow them.

It requires appropriate tools including special tongs to place the terminals and photovoltaic connectors. If we know your needs, we can send ready-made cables according to your instructions.

THE INSTALLATION OF PHOTOVOLTAIC PANELS

The installation of photovoltaic panel on a roof varies depending on the accessories: installation on metal rails, on an insulating membrane or an impermeable layer. The watersheds can be zinc, lead or plastic.

INSTALL SOLAR PANELS: DIFFERENT DEVICES

The photovoltaic panels are mainly installed on the roofs (inclination of 30 to 35 degrees and recommended south orientation) and their fixing systems are different and influence the ease of installation, the necessary accessories, the duration of intervention, the degree of competence and the tools necessary for placement.

The costs vary considerably depending on whether the rails are placed on metal rails on the tiles, if the cover has to be removed to leave a hole in the panels, or if it is necessary to waterproof the roof before placing the panels.

The installation of solar panels allows the production of electricity . The choice of solar panels , depending on specific criteria , requires a prior study and a precise definition of the needs of the installation and / or the home.

HOW TO DO IT

To install solar panels , you need to feel good working at height and have knowledge about covers and tightness to avoid water leaks. You will need to ensure the correct fixing of the panels, as they must withstand unfavorable weather conditions, place the clamps, rails and slats.

Read the assembly instructions carefully to understand the connection diagram. Work safely when handling electrical components, for example when installing the inverter, connecting the electrical cables and fixing them on the wall.

Knowing how to use a drill and a circular saw are also some of the requirements. Depending on the solar panels and their fixing mode, you may have to install a support under the system or a sealing system (stapled in the first case and screwed in the second). A good physical form is also necessary to work on a roof and manipulate photovoltaic panels.

Realization time

6 hours minimum (depending on installation)

Number of people

2 people

STEPS FOR REALIZATION

1. Prepare the roof where the solar panels go;
2. Place the watersheds (lower and lateral);
3. Place a sealing or waterproofing system;
4. Place the rails and panels;
5. Connect the inverter to the network.

TOOLS AND MATERIALS

- Screwdriver and / or drill-screwdriver with different tips
- circular saw
- Extension
- Subway
- Scaffold
- Professional ladder (depending on coverage)
- Pipe wrenches or tips
- Screws
- Gaskets and other hardware parts to fix the rails
- Water drops and profiles
- Insulating strip
- Slats and battens
- Silicone and silicone gun
- Electrical cable and cable tubes (depending on configuration)

- Protection gloves
- Fall arrest system
- Safety glasses
- Safety shoes

INSTALL SOLAR PANELS

The panels can be placed on the floor, on a wall or on a roof . The fastening system will vary according to its location.

If your roof has tiles, the rails can be placed on the tiles, removing only the tiles located at the anchor points of the rails, on which the plates are held.

Another possibility is that, taking advantage of the inclination of the tiles, the panels are placed directly on the roof, with a good waterproofing and replacing the tiles with the panels that, in that case, will have a double function: as part of installation and as protector of the roof or roof.

1. REMOVE THE COVER IN THE INSTALLATION AREA OF THE SOLAR PANELS

As we have seen, two distinct cases are distinguished. If the roof is covered with tiles or slate, you will need to remove a part. You have to start by defining the exact location where you want to install the solar panels.

If you are going to use the panels as a "roof", you must remove the tiles from a surface slightly higher than that of the solar panels. Once this operation is done, a hole will remain in your roof and in that hole is where the photovoltaic panels will be housed after a meticulous preparation.

Wear protective gloves because the tiles are abrasive and can cause a hand injury. Any intervention on the roof must be carried out with an anti-fall system (safety net or harness). If you remove the tiles near the ridge, be careful not to break the header or longitudinal tiles (do not walk on them and lift the tiles gently).

If your roof is not reinforced (truss frame), do not walk in the middle of the slats or on the roof membrane or insulating layer, you run the risk of falling.

2. PUT THE LOWER AND LATERAL DRAINS

The gutters are connecting pieces between the roof tiles and the solar panels and their fixings. They provide a perfect seal . There are waterslides adapted to each model of tiles, of different shapes and colors. For blackboards, it is a slate colored sheet or a piece of zinc.

The gutters are fixed with screws to the modified or added slats where the solar panels will be received. The use of a screwdriver drill is necessary, as well as a meter to calculate the exact measurements . The clamps are also installed with screws.

If necessary, include a ribbon along the panel and behind the first row of shingles to gain height and provide a better base for the watersheds. This can be made of plastic and come with the solar kit, or be a band of asphalt fabric that must be placed on the ribbon and the first row of tiles. This insulating band (of variable width) is glued with silicone applied with an applicator gun . The asphalt cloth tape is molded with the hands to adapt it to the shape of the tiles . The plates and slats are cut with a circular saw .

3. PLACE THE INSULATING MEMBRANE OR A WATERPROOFING SYSTEM

Insulation membrane placement

During the installation of a roof, an insulating membrane must be placed . This is a technical plastic layer that has the function of creating a waterproof but breathable barrier. It is recommended to leave an air layer between the insulating membrane and the waterproofing layer in order to avoid the appearance of mold .

The membrane is usually stapled to the wooden roof structure. If this is impossible and the membrane has to be placed directly on the insulator , then select a membrane of high water vapor permeability, HPV , which allows the passage of steam and the risks of condensation on the insulator .

If condensation occurs in the insulator, it becomes less effective and can even rot. The membrane joints must be joined with the adhesives recommended by the manufacturer and according to their indications.

Install a sealing system

Depending on the installation, solar panels can also be placed on a waterproofing system composed of flexible plastic parts that must be installed on the roof (a rigid base consisting of wooden or aluminum slats is necessary). This sealing system is screwed onto the slats in the area where the solar panels can be located.

4. PUT THE METAL PROFILES AND PHOTOVOLTAIC PANELS

The profiles must be firmly fixed to the beams with the help of clamps (use a suitable screwdriver and screws), since the solar panels are heavy and offer great wind resistance. In this case it is not necessary to remove the roof tiles or roofing because the metal profile structure is above the roof.

It is especially important to respect the maximum separation between two profiles recommended by the manufacturer, as well as the maximum distance between two fixings . If the profiles are not long enough or have to cross, there are several connection pieces.

It should be noted that some solar panels can be placed vertically, that is, high and not wide. In this case it will be necessary to lay the rails vertically. The panels are then fixed on the profiles with the help of clamps or screws . Once all the panels are installed, you can fix the upper watershed and replace the missing tiles or parts.

If you are going to use a waterproofing system , the metal profiles can be installed directly on top .

5. Connect the panels to the inverter

The inverter is an electronic device that transforms the unstable direct current at the output of the photovoltaic panels into an alternating current of 220V and a frequency of 50 Hz (frequency of the conventional mains). This inverter is connected on the one hand to the panels and on the other hand to the consumption meter , which is located at the head of all the individual electricity production facilities connected to the grid . It is important for this stage to know how to read an electrical scheme.

Some panels must be connected in series and others in parallel in order to obtain the voltage and intensity values around the nominal values provided by the inverter, in order to preserve their longevity.

Two panels of identical characteristics are connected in series if the positive (+) pole of one is connected to the negative (-) pole of another; Your tension will add. If each panel produces a voltage of 12V, the two series panels produce a voltage of 24V.

If you join the poles of the same polarity between them, the panels are connected in parallel. This assembly produces a voltage of 12 V, but the intensity produced is the sum of the intensity produced by each panel . Attention! If you join series and parallel panels, the tensions of each group must be the same to preserve the panels and maximize their production.

COMPONENTS OF A SIMPLE PHOTOVOLTAIC SYSTEM

A solar system components must be sized according to local requirements and conditions. Our solar calculator can help.

The teams must have the certifications according to current regulations. In Peru they are the NTP Peruvian Technical Standards and the IEC standards (where Peru is a member). Frequently the NEC, IEEE / ANSI (US) or similar standards are used, although these are not always compatible.

Standards are not guaranteed for products that are purchased in some famous markets, where you can find those that did not pass the rankings. Do not be guided only by prices!

- A simple traditional system (DC coupling) of low voltage consists of

- solar panels (or modules), to be placed on a ceiling mount or on a pole,
- a controller that regulates the photovoltaic current and protects the batteries,
- batteries suitable for cyclic use to charge and discharge them daily,
- a sine wave inverter (if 220V is required). Be careful, cheap inverters that do not produce a pure sine wave affect the life of many devices and can break them (eg compressors in refrigerators),
- fuses (breakers),
- cables and connectors,
- the distribution of energy with pass boxes, etc.,
- a ground connection and,

SIZING THE SOLAR PV SYSTEM

HOW TO CALCULATE THE SIZE OF A PHOTOVOLTAIC SYSTEM

Not all photovoltaic systems are the same, as the energy requirements of the different areas where energy is used are not the same. The methods of consumption differ from one case to another and taking this variable into account is important for choosing the right size of the photovoltaic system and the right installation solution.

The demand for electricity, in terms of quantity and in terms of time slots of use, can be very different depending on whether it is a small domestic plant, a medium company plant or a large commercial plant.

To calculate the right size of a photovoltaic system you will first have to ask yourself what you install it for.

Do you install it mainly for self-consumption or to produce and sell energy on the network? Based on the answer to this question, the size of the plant may be more or less proportional to its consumption.

If the plant is mainly built to meet its own energy needs then it will be worthwhile to size the plant so as to guarantee a production slightly higher than its annual consumption. In this way you will have the best cost-benefit ratio (if we look only at the benefits in "absolute value", obviously, the larger a plant will be, the more it will provide economic revenues).

The right size of a domestic photovoltaic system will be proportional to the annual consumption of the house to which it is connected, to the installation area (north, central or south Italy) and to other specific criteria such as: orientation and inclination of the modules, shading, exposure, etc. ..

CALCULATE SIZE OF PHOTOVOLTAIC SYSTEM

If I have a consumption of 2,500 kwh / year and I live in an area that produces on average 1,200 kwh / year a 3 Kw plant will be enough to cover my entire needs and to have a good amount of energy fed into the grid that will help me better amortize the construction costs of the plant. Not all the Kw consumed will be produced by my plant, but the amount of energy produced by the plant will be partly self-consumed immediately and partly fed into the electricity grid and enhanced by the exchange on the spot or by the dedicated withdrawal (the dedicated withdrawal is the sale of the 'energy to the electricity manager).

In general, on-site exchange is more convenient if one's consumption is close to the estimated production of the photovoltaic system.

Instead, the dedicated withdrawal (ie the sale of energy to the grid operator) is worthwhile if the size of the photovoltaic system allows a production much greater than one's own needs.

For domestic systems at the service of a family of 3-5 people, it is advisable to place a system of 3 to 6 Kw (in an on-site exchange system). If you want to build a larger size system, you can evaluate the option of dedicated withdrawal with the installer (ie sales with "guaranteed minimum prices").

The general rule is that the more the loads are distributed constantly throughout the day or week, the more solar energy produced by your photovoltaic system will be easily exploited through self-consumption. All the energy that can be self-consumed at the moment is always a source of savings for the user.

INSTALLATION PROCESS

THE PLACE

With the exception of solar panels, all equipment is installed in a weather-protected, dry, cool, ventilated and dust-free place. Normally everything is placed in an unused corner inside the house or under other protection.

Animals and children must not have access. Do not forget occasional events such as earthquakes, floods, etc.

Solar systems can cause fires and batteries are toxic, contain acids and a power outage can cause severe burns.

FACILITIES AT HEIGHTS

Two factors that influence the proper functioning of the devices depend on the installation altitude and should be considered:

The air density is reduced and decreases the capacity of cooling equipment. They can heat up quickly, including burning. If the temperature is excessive, good equipment reduces the power and goes out, bad equipment is broken,

The lower atmospheric pressure affects the insulation between the conductors. In height, dielectric strength is reduced. Electric leakage may occur or sparks jump, damaging controllers and inverters. This also affects other electronic equipment, for example computers frequently die at heights.

Up to 2000 meters above sea level there are no problems. Quality electronic equipment is certified at least up to 2000 meters, good up to 3000 meters, but very few up to 4000 meters or more (see standard on Classification of environmental conditions IEC 60721-3-4).

Getting certified equipment for the expected height is important to ensure proper operation and to maintain the manufacturer's warranty. If this is not possible, the equipment must be oversized so that they do not work under maximum capacity.

Putting them in a cool, dry and well ventilated place is of the utmost importance. Our highest installation is at 4867 meters and it works without problems.

TYPE OF ELECTRICITY

In small photovoltaic systems, direct current (DC or DC, short for English) is used. It has a different behavior than the common current of 220V AC. For example, AC fuses fuse under DC current with other values. Do not make a mistake with the positive pole and the negative pole.

Small systems up to 500Wp with several crystalline modules can reach a voltage of approximately 72V that does not represent a danger to life (but is enough for a strong scare).

With more modules in series, and with some panels of the third generation (for example the amorphous), the voltage is higher and dangerous. Proper precautions are necessary! With a 220V inverter, the same attention is required as an installation connected to the network.

CABLES AND CONNECTORS

Cables conduct the current, and improperly connected or damaged, can cause loss and damage. Many of the problems do not arise at the beginning, they come with time.

In fixed photovoltaic installations, very flexible cables are not used (for example, those used for welding). The connections must be good and proper terminals must be used.

There are always losses of electricity in the cables. In low voltage systems, they should be considerably fatter than in normal installations (think of thick battery cables in cars).

Although a maximum loss of 3% is allowed, it is better to keep a lower value. In case of doubt, you should always choose the thickest cable. There are several sites on the internet that allow you to calculate the dimensions of the cables . To optimize an installation it is: keep the cables as short as possible, use a diameter of the cables (measured in mm 2 or AWG) sufficient to avoid losses and that the cable is heated.

CONNECTIONS

They are of high importance and frequently cause failures. To prevent corrosion and ensure good contact over years, flexible cable requires terminals. Clemas or torsion connectors are required to ensure a reliable connection.

Rain and dew moisture in the morning cause water to stick to the wires, to prevent it from entering electronic equipment, most connections are placed below the box so that water cannot infiltrate. If this is not the case, you have to put a loop so that the water is outside. Many teams also have rubber caps to inhibit water (and animals like spiders) from entering.

To pass the cables of the solar panels outside through the wall or a window, the same principle applies: keep the water outside. Tipping the hole out and looping prevents water from entering. An additional insulation is advantageous.

Distribution box with good Wigo connectors: Of course, the cables must be fixed so that they are protected, that they do not move excessively with the wind and that they are out of the way without putting a danger.

Cable ties that resist ultraviolet radiation are helpful. Special photovoltaic cable and vulcanized cable can be placed without special protection with staples, loose cable requires a pipe. In the weather it is very important to protect normal cable against ultraviolet radiation that in a few years can destroy the insulator, first it gets hard, then it breaks.

SOLAR PANELS

Module Management

There is no need to put weight on the modules and for transport they must be protected against any blow.

Although quality modules withstand the weight of a man without the glass breaking, the very small connections inside the modules are very sensitive and any deformation can damage them that reduces efficiency.

ORIENTATION

The orientation of the solar panels south of the equator is to the north. The maximum energy is captured if the inclination coincides with the geographical altitude of the installation site. This, for example, for Arequipa is 16° and for Piura 5°

In reality, the panel is more inclined, between approximately 20° and 25°, accepting a reduction in annual yield. Namely:

To adequately make better winter performance when there is less sun and more light is needed, the inclination of the panels is increased to capture more energy in this period.

Dirty panels must be clean so as not to lose energy. Dirt reduces the performance of solar panels. In some places, for example next to an unpaved road, it can be very strong. The more inclined, the less they get dirty. In rainy areas it is used in addition to 'self-cleaning': a normal rain washes a panel automatically if the inclination is more than 20 °.

Special conditions influence the orientation. The best panel performance can be adapted to the hours of maximum energy use during the day. If mountains cover the sun in the early hours of the morning, turning the panels can optimize performance. In larger systems, part of the panels can be oriented to the east and another to the west to improve the distribution of the irradiance curve.

PANEL MOUNT

The panels can be mounted anywhere (avoiding shadows), it is normal on a roof or a pole. The structures must be strong enough to resist the strongest winds that can occur in the area,

Avoiding theft can be important. There are several solutions, for example special screws that require a special wrench (similar to car wheel theft protection) or the modules can be encapsulated.

VENTILATION OF SOLAR MODULES

The colder, the better the performance, because at elevated temperatures the panels produce less (except organic panels) Solar panels are heated not only by ambient temperature, they produce their own heat. Much of the radiation not converted into electricity is transformed into heat!

Without ventilation, panels can easily be heated to more than 50 ° C and thus lose more than 10% energy. Poor quality modules can lose up to double.

It is of great importance to install the panels in such a way that heat does not accumulate in leaving spaces.

BATTERY

Lead - based batteries require a cool place to keep much as possible a temperature between 20 and 25 ° C. A lower temperature (but not below the freezing point) does not damage the battery, it only reduces the capacity. A higher temperature greatly reduces its life (for example at 35 ° C, life is already half).

Batteries must be installed in a ventilated place, because all lead-based batteries always produce hydrogen, an explosive gas. In VRLA type batteries (those of Gel and AGM) the gases are combined internally and if there is no overload, gases cannot escape.

No battery is completely sealed, they have valves in case of an overload. Liquid batteries, sealed or not, do not have this recombination and the amount of gases that escape can be much greater.

The battery connections must have a special grease (alternatively petroleum jelly) to avoid sulfurization and corrosion of the poles.

DC FUSES

To secure the system against overloads and power outages, fuses are always placed between batteries and controller / consumers. Its capacity is determined by taking the maximum current plus 15 to 20%.

There are fuses for high DC type currents, but these are sometimes difficult to obtain in the local market. Alternatives are fuses that are used for car sound amplifiers, or fuses that are used in the marine.

Common fuses can be installed for use in AC systems but be aware that these thermo-magnetic fuses trip under a DC current with different values. Good manufacturers give the values for DC in the technical specifications.

Note: Although modern fuses or breakers can occasionally be used as a switch, they cannot withstand a high amount of turning on or off. The cause is the material of the contacts that is optimized to reduce losses. If there is a need to connect and disconnect the batteries frequently, a separate switch must be installed.

THE CONTROLLER

The controller converts the DC current of the panels to a current suitable for the batteries. It protects them against overload and excessive discharge. Without this protection, batteries can get hotter than normal, lose liquid and produce excessive explosive gases: they can explode. An overload also reduces life. On the other side it is important that the batteries are fully charged to avoid rapid sulfurization.

Different batteries require different ways to charge them.

Charge voltage (volts) and current (amperes) vary and depend on state of charge of the batteries,

- Battery Type. Liquid batteries (sealed or not) and AGM require a different voltage than Gel batteries,
- Temperature. The hotter the environment, the more the voltage should be reduced to avoid gasification,
- Age of the batteries.

Good controllers fit different types of batteries and have a temperature sensor.

These factors can make a difference of several years over the life of your battery. A good controller disconnects consumers, once the battery charge drops to a certain level. Therefore, in normal installations, the output of electricity for the use of electricity is from the controller.

Sometimes, for example, if an inverter that exceeds the limit of the controller current is connected, consumers can be connected directly to the battery. In this case, make sure that the battery is automatically disconnected when it reaches its discharge limit. Good investors have this built-in control, but not cheap ones. We have seen several batteries killed in a short time by this careless.

DRIVER INSTALLATION

Drivers frequently come for different battery voltages. Many handle 12 and 24 volt batteries, some up to 48 volts and more. Most controllers automatically adjust to the nominal voltage of the batteries. Therefore, it is important that the controller is connected first to the batteries and only later to the solar panels.

In some controllers there are small switches ('dip switch') that must be adjusted according to the voltage and type of the batteries, there are others that have connection bridges that must be placed according to the requirements.

Always take care to connect the positive and negative poles correctly. Drivers that do not have protection against a bad connection can be damaged immediately.

If possible, terminals must be used to connect the cables. This ensures a good connection and prevents oxidation over the time that causes losses and is sometimes a reason for sparks and reduced redness.

In more powerful controllers, the temperature sensor is separated to place it directly to the batteries. If there are several batteries installed, a place is sought among the batteries that reflect the maximum temperature of the batteries and not that of the environment. Sometimes the sensor cable is very long, it should not be cut, because the cable is part of the system.

THE INVESTOR

We strongly recommend getting a sine wave inverter, because cheap non-sinusoidal or semi-sinusoidal ones shorten life and can damage sensitive devices such as refrigerator compressors and other electronics. They cost more, but the additional cost is justified. Quality inverters also come with all the protection for batteries, against bad connections, overload and contain fuses that increase safety.

THE CONNECTION

An inverter is connected to the controller if the output of the controller can maintain the amperes that the inverter requires for maximum power. Frequently this is not the case and is placed directly to the battery in parallel to the controller. For this, battery connectors are used that allow the connection of both cables in a clean way to ensure good contacts. The 220V connections of the inverter output are made the same as a common 220V installation. Do not forget the fuses!

CHAPTER EIGHT

TIPS FOR USERS OF PHOTOVOLTAIC SOLAR ENERGY INSTALLATIONS:

Do not connect high-power equipment to the photovoltaic system that has not been considered in the design, without consulting specialists, since an overload due to excessive consumption can cause a malfunction.

Do not connect equipment with higher power than the DC / AC inverter, as this overload may damage it, especially when the inverters are not of quality

Remember that all devices with motor have at least a starting power 3 times higher than the power of the device. It must be taken into account to know if our investor will accept it.

No modifications should be made to the installation, since the installation has been specifically sized for the use that was initially established.

Do not use incandescent lamps. It is advisable to use LED lamps or, failing that, low consumption.

It is not convenient to use devices with electrical resistors type: braziers, radiators, heaters, electric water heaters, etc. Its consumption is excessive. It will be more appropriate to use another energy source for heating.

Always remember that in photovoltaic solar energy systems, as energy is limited, it becomes much more necessary. Therefore, do not keep lights or equipment on unnecessarily.

Check weekly the charge regulator indicators, which indicate its operating status, and verify that it has a regular production.

Under no circumstances should the air outlet of the inverter be plugged, since if this is the case, it is prevented from cooling and could lead to a malfunction.

If the inverter is protected, either by overvoltage or overcurrent, and turns off when we subject it to a load greater than what it admits. We must not restart it, after a few minutes the inverter will automatically reset.

Urgently report any breakdowns to the technical service.

Check that the appearance of new shadows (vegetation, new constructions) may reduce the electrical production capacity of the installation.

Try to reduce the electrical consumption of the site so as not to overstress the battery, thus prolonging its useful life.

Once a year check the water level of the batteries of your solar energy installation. Do not use, instead of distilled water to refill the accumulation battery, river water, boiled water or any other type than recommended, as this damages the life of the accumulation battery. If the battery level is refilled, it should be done using a plastic or glass funnel (in no case use metal containers).

Battery charge:

Below is a table that reflects the percentage of charge based on the voltage produced by the battery.

In case of needing the replacement of protection elements (fuses, circuit breakers, differential, etc ...), special care must be taken in the disconnection, first of all the fuse holder base belonging to the panels must be opened and secondly the one corresponding to the accumulators or batteries. It is preferable to call a specialist for this operation.

At least one annual periodic inspection by the installation company is recommended.

It is recommended to clean the solar panels with soap and water once a year.

Nothing should be put on top of the batteries. Do not manipulate their terminals with your hands. Do not leave them within reach of children. The space for solar energy batteries should be very well aerated, since most solar batteries emit gases.

It is very important that the solar energy installation has the appropriate protections, so it is recommended that if you doubt the safety of your installation contact a solar energy specialist to advise you.

FURTHER TIPS ON A PHOTOVOLTAIC INSTALLATION

- For a fixed installation, the decisive criterion is that of the installed power in relation to the space you have.
- The solar panels high efficiency produces more per square meter and, therefore, are more expensive for its most advanced technology and manufacturing quality.
- To compare the solar panels, observe the maximum power per square meter (Wp / m^2).
- Apply the calculations to achieve a specific production objective taking into account all the parameters of your installation. The formula we offer allows you to evaluate the annual

production, the peaks of summer production compensate for the low winter production.
- Anticipate a significant margin if the panels will be used to power appliances throughout the year, so you will never run out of electricity.
- Remember that the choice of location is paramount. Make sure there is no shadow on the panels. Remember that trees grow and do not forget to periodically clean the surface of your panels after installation to ensure optimal electricity production.

HOW TO SAVE ENERGY

Roof condition: To take advantage of the high durability of the panels, make sure that the condition of your roof is in good condition and / or make the necessary arrangements before installing the panels since it can be expensive to uninstall and reinstall panels for a roof arrangement.

Get advice with a professional! The angle of your roof is also important. A solar consultant can take measurements and inform you how well your panels could perform on your roof.

Cleaning: Solar panels require very little maintenance. If dust or leaves accumulate on the panels they can be easily washed with hose. Beyond that, they practically do not require other maintenance. Uncleaned panels can lose more than 5% of their generation potential!

Maintenance: The panels are usually guaranteed for 25 years and the inverter 5 to 10 years. The inverter usually needs to be replaced after 10-15 years, as well as the batteries. It is always advisable to consult a solar consultant to examine the state of your system every 4-7 years.

Property valuation: Like any improvement over a house, a solar system increases the value of your home, that is, it works as an asset that adds value to your home in case of sale.

HOW TO CHOOSE FUSES AND OTHER MODULAR CONTROLS

Differentials and circuit breakers or circuit breakers are your guardian angels. However, an electrical panel can contain many different types of modules, which can be old as fuses, or economical as limiters to take advantage of the nightly rate.

Circuit breaker, or circuit breaker, and overload: how to manage it

The price of your rate depends on the contracted power. If you consume more than the maximum power, the leads will jump. That is, the calibrated power control switch (ICP) cuts off the power input supply, if the contracted with the power company is exceeded. Likewise, the IGA (Automatic General Switch) is disconnected when the power is greater than that allowed by the installation, avoiding overloads and short circuits.

To solve this problem, there are two options:

- increase the power of the meter and pay a more expensive rate;
- Install a power rationalizer that cuts the least priority circuits gradually.

The power rationalizer automatically manages household electrical appliances. Compare the power consumed in relation to the power set and cut the power of the less priority appliances, if necessary.

These devices are usually electric radiators or heaters, which consume a lot and which we can do without for a short period. If the power consumed decreases, the contactor restores circuit power and prevents the limiter from jumping.

Priorities can be defined within the circuits; give privilege, for example, to a heater in front of a radiator. It is the ideal solution for a holiday home where everything is electric and it would not be reasonable to pay a higher fee for a few cold days a year.

HOW TO TAKE ADVANTAGE OF THE NIGHT RATE OR DISCRIMINATED RATES

With a nightly rate or with the discriminated rate the electricity consumed at night or in the chosen time slots is cheaper. To achieve maximum savings, a power rationalizer can be installed in the electrical panel. The priorities and the cutting order of the circuits are selected. There are three modes of operation: in automatic march, according to your needs. For example, the rationalizer allows you to use the hair dryer that consumes a lot of power, and in the meantime cut off the supply to the electric radiator to reduce consumption by not adding up to the consumption of both at the same time. Thus, the appliances are powered by cheaper electricity; in forced march, the circuit is powered.

It is essential if your heater needs to heat the cold water while you shower, and at the same time you put a washing machine, but it will be more expensive; the shutdown, allows to prevent any electrical consumption, ensuring a substantial saving.

FUSES AND AUTOMATIC: EVERYTHING YOU NEED TO KNOW

Fuses and automatic

The fuses protect the phase of the circuits and blow if they are crossed by a current that exceeds a certain value . As with circuit breakers or switches, there are different gauges depending on the use . However, it is not possible to make all installations with fuses. In the case of a VMC (Controlled Mechanical Ventilation), for example, a circuit breaker must be used.

The fuses are installed in the automatic devices , which cut the phase when they are in the open position, which provides additional safety when it interferes with the corresponding electrical circuit.

Fuses have various sizes and functions to avoid confusion, which would be very dangerous. A 20 A fuse on a line that must be protected at 10 A would not do its job well. It would let a current of 15 A pass that could heat the cables and cause a chain fire.

Before it was difficult to know if a fuse had jumped. Today, some automatic devices have an indicator. In one of the fuse terminals there is a pickup that pops up when the fuse blows. It is ingenious and avoids having to check the fuses one by one with a multi-meter.

The connection or grounding distributor

CONNECTION OR GROUNDING DISTRIBUTOR

The grounding distributor is a distribution block to which they must be connected: the wires of each circuit connected to the ground, whose section is equal to the phase and neutral of said circuit.

The cables of the equipotential connection of the dwelling, that earthen all the metallic conduits that do not have to carry current, that enter the house; the cables of the equipotential connection of the baths, which earthen the entire metal mass; the cable connected to the grounding of the house, whose section is equal to that of the phase and to the neutral that comes out of the electric meter: a connector must be provided for each cable, reserving 20% for future needs.

INCREASE PROTECTION WITH A SURGE LIMITER

Surge Limiter

Lightning strikes or strong power surges can have serious consequences on your installation, since they can generate voltages of hundreds of thousands of volts, against which the classic protections are not effective.

In areas exposed to lightning, or voltage surges of more than 10%, the standard requires the installation of an overvoltage limiter.

The modular limiter is placed at the beginning of the installation. A pilot indicates if it is operational or if the protection cartridge needs to be changed. The limiter must be connected to the ground and placed as close as possible to the meter and the ground for maximum efficiency.

WHEN A REMOTE SWITCH IS NEEDED

Remote switch

In an electrical installation, a bulb can be activated from two switches thanks to what is known as a switched installation.

For anything else, such as controlling a group of lamps, it is necessary to use a remote switch or remote switch. It is a module that is placed in the electrical panel.

It is usually used to control several lamps. That is, it is connected to the set of buttons that activate the lighting and light points.

When it receives a pulse from one of the switches, it changes state, turning the light on or off.

In addition to the unipolar, bipolar or three-phase versions, we can also find a time switch or "automatic ladder", which turns off the light after a while. It is ideal for common areas of passage or if you have your head in the clouds. In their silent version, they are quite discreet, even placed in the junction box or register of a room or near a room.

HOW TO ACTIVATE A LAMP FROM A LIGHT SWITCH OR MECHANISM

Light Switch or Mechanism

You had already thought about it, but the manufacturers did it before you.

There are switching modules or pushbutton switches to control the illumination of a light point, from one point to another, or from a remote control switch.

It may be practical , for example, in a professional premises, to centrally control certain points of light.

HOW TO CONNECT A GENERATOR

To be able to carry electricity produced by an alternative source, you have to turn to electric inverters , which have two modes of operation:

with normal operation, they use the network as a source of energy;

in case of a cut, they send a signal to activate the emergency sources to power the circuits to which they are connected.

Attention: its implementation is complex . It must be ensured that the generator is protected, that you will not forward electricity to the power grid (which is dangerous) and that the power consumed in the circuits to be fed does not exceed the capabilities of your generator.

HOW TO CALCULATE THE POWER AND SIZE OF A SOLAR PANEL

Do you want to have electricity in an isolated booth or to go camping? In these cases the installation of a solar panel is evident, you just have to calculate the necessary power and the size of the solar panel. Here are the calculation formulas that you must apply.

The amount of electrical energy is expressed in watts hour or Wh. It is simply to multiply the power in watts, by the time during which it is consumed or charged. For example, a 1000-watt radiator on for one hour will have consumed 1000 Wh or 1 kWh. The kWh is the standard unit of measure that can be found on your energy supplier's bills.

To estimate the energy requirement of each of your devices, simply multiply the nominal power in watts by the usage time in hours. You will obtain in Wh the minimum amount of energy to produce to power your devices.

It should be noted that the power produced by a solar panel is greater at noon than at the beginning of the morning or late in the afternoon, you will surely need to use a battery to store electricity and a converter to power your devices.

WHAT FACTORS INFLUENCE THE ELECTRICITY PRODUCTION OF A PHOTOVOLTAIC PANEL

The energy production of a photovoltaic panel depends on both the quality of the equipment , its environment and the circuit it feeds.

THE POWER OF THE SOLAR PANEL

Rigid solar panel

Expressed in kWp or kilowatt-peak, it is the electrical power produced under ideal conditions . According to technology, a square meter of solar panel can produce 60 to 150 Wp . Once the peak power is known, you can calculate the total area of solar panels that you will need.

THE SOLAR RADIATION OF THE AREA

Less than kilowatt hours (kWh) are produced per kW peak (kWp) in Galicia than in Malaga. This magnitude, the sun, is interesting because it allows to evaluate the annual production taking into account the place where the panel is installed from the peak power of the panels .

INSTALLATION PERFORMANCE

If the panel is not oriented to the south and at an angle of 30-35 ° from the horizontal, it produces less electricity than it is capable, since the sunny surface is smaller. Taking this performance into account allows you to improve the installation.

HOW TO CALCULATE THE ENERGY PRODUCED

Panels kit

The total energy produced E_p represents the number of kilowatt hours (kWh) produced by the panel in one year. It is calculated using the following formula:

Ep = r * Ens * PC

But what interests you is to calculate the peak power of the panel you want to buy to cover your electricity needs or sell a certain amount of energy to an electricity supplier.

By transforming the above formula, you get:

$P\_C \geq (E\_p * F\_CO) / E\_ns$

with F_CO = 1 / r with a correction coefficient to compensate for performance losses

PANEL PERFORMANCE

The performance of a panel depends on its components:

6% for amorphous silicon;

15% for polycrystalline silicon;

16 to 24% for monocrystalline silicim.

ABBREVIATIONS USED

kWh : kilowatt hour

kWp : kilowatt-peak

E_p : total energy produced

r : performance

Ens : sun

P_c : peak power

F_co : corrective factor

HOW TO CHOOSE THE AUTOMATIC OF YOUR ELECTRICAL PANEL

Any current greater than 50 mA is dangerous. The role of the automatic and differential circuit breakers is to cut the power supply in case of overload or current leakage greater than 30 mA, and thus reduce any electrical risk caused by an insulation failure.

OPERATION OF A DIFFERENTIAL SWITCH

The differential circuit breakers comparing the intensity of the stream entering the installation with departure, if the difference exceeds a value we call sensitivity , expressed in milliamps (mA), the differential stops.

The leakage of current, which is the one that is lost, may have circulated through the metal housing of a device for grounding, or worse, through a person who has touched the conductor. Thus, it is important to have, on the one hand, differentials at full capacity and, on the other hand, a grounding in conditions.

WHY INSTALL A DIFFERENTIAL CIRCUIT BREAKER

The circuit breaker at the meter level has a sensitivity of 500 mA , ten times higher than the 50 mA limit we have already talked about. This circuit breaker does not offer enough protection for people. Therefore, 30 mA differentials are necessary, which are also called high sensitivity differential devices .

In addition, the meter differential protects the installation as a whole . In case of any problem, all the electric current is cut off, which makes it difficult to find the fault, in addition to the main devices can stop working. It would be a shame if the refrigerator does not receive power because of an insulation failure in an electric coffee machine.

Modern differential technologies provide solutions to the problems caused by all electronic devices we have at home.

DIFFERENTIAL SWITCH OR CIRCUIT BREAKER

The switch is the simplest of the two, since it only offers the differential protection that we have already talked about. The circuit breakers, in addition to their differential function, protect the circuit from power surges and short circuits. A differential switch is designed to be installed above several differential circuit breakers , which feed the non-specialized sockets and lighting circuits. It is economical, since a single differential protects several circuits.

The differential circuit breakers are reserved for the lines to which the specialized sockets of the devices that have to have current yes or yes are connected. By combining the functions of the differential switch and differential circuit breaker, we save the installation in the panel.

This type of protection is ideal for devices that must receive current for as long as possible , despite sporadic insulation failures that arise at other points in the electrical circuit. Differential circuit breakers have always been used to protect the circuits that supply computer or cold storage equipment. For current equipment, it is very likely that you need one of 20 A.

MEANING OF THE CALIBER OF A DIFFERENTIAL SWITCH

The value of the gauge expressed in amps (A), corresponds to the maximum intensity that the switch can go through without damaging it.

TYPES OF SWITCHES AND CIRCUIT BREAKERS

AC type

It is the most common. Those of type AC are only activated for leaks in alternating current . This excludes, therefore, transient leaks caused by lightning or the charging of a capacitor. The AC type are used for common installations: lamps, small appliances, etc.

Type A

Type A shields against direct current overloads, in addition to AC type protection. This current is present in certain switched sources that transform alternating current into direct current. These type A differentials are used for the electrical installation of washing machines or ceramic hobs or inductors.

Type Hpi, HI or Si

The types Hpi, Hi or Si, according to the name of the manufacturer, are high immunity differentials , designed to avoid untimely activations. They are used to individually protect the power points of devices that cannot suffer power outages , such as refrigerators or computer equipment.

DIFFERENTIALS: RULES AND PRICES

There are different price ranges. It must be ensured that they respect IRAM 2169, which comes from the European standard EN 60 898. The reference marks are the most expensive, but are internationally recognized for their reliability.

Within the same range, type A differentials are 20% more expensive than those of type AC. Hpi, meanwhile, are three times more expensive. Those of type A can be replaced by those of type AC or Hpi instead of the rest. Of course, provided that the rule authorizes it.

HOW TO DETERMINE THE NUMBER OF DIFFERENTIALS

The minimum number of differential switches that must be purchased depends on the surface of your home. To get an idea, you will need:

2 if the surface is less than 35 m2;

3 if the surface is less than 100 m2;

4 if the surface has more than 100 m2.

In any case, at least one of type A with a nominal intensity of 40 amps, and the rest of type AC with the same nominal intensity will be necessary. The nominal intensity can rise to 63A if you have radiators with a power greater than 8 kW. If you have devices whose power you want to preserve, you can complete it with as many Hpi circuit breakers as sensitive devices you have (computers, refrigerators, etc.).

ESSENTIAL POINTS FOR A GOOD CHOICE

The number of differentials depends on the surface of your house and you will need at least one of type A and the rest may be type AC, which is cheaper.

With a nominal intensity of 40 A it will be sufficient, except if you have very powerful electric heating, in which case one of 63 A will be used.

The specialized sockets that give power to sensitive devices can be protected with differential Hpi circuit breakers, with their consequent inversion.

ONE LAST TIP TO CHOOSE AUTOMATIC

The choice of the circuit breaker depends on the circuit to be protected and the regulations in force. Consult the UNE-EN 60898 and UNE-EN 60947-2 standards before installing or activating your electrical circuit.

Investing in protection is essential, I advise you to choose good quality protection systems.

Prices may vary from one brand to another, but only good quality circuit breakers will fulfill their function for years.

They will jump automatically when the nominal current is exceeded and will be reactivated alone instead of burning like simple fuses .

They are designed for simple use and installation , which will save you time and avoid headaches!

How to choose electric wires and cables

The section of electrical wires and cables varies according to intensity, installation and equipment. 3G 1.5 mm 2 wire, monofilament wire, copper wire, coaxial wire, RJ45 connector or music wire: large family! In order not to confuse the phase, the neutral and the earth, we leave you the following guide.

DIFFERENCE BETWEEN A WIRE AND AN ELECTRIC WIRE

Electric wire

An electric wire serves to transport electricity. It is formed by a core, symbolized by the conductive material, which can be mono or multifilament depending on its section and its purpose. The core is usually copper, but it can also be nickel. It is covered by an insulator, which is usually plastic.

Electric cable

A cable is a set consisting of several electrical wires, grouped within a sheath, or protective hose, of plastic. Both its number and its diameter vary depending on the use of the cable in question. There is a European CPR Standard for electrical cables, mandatory.

CRITERIA THAT DETERMINE THE SECTION AND COLORS OF THE CONDUCTIVE WIRES

The section of an electric wire

The section is expressed in mm 2 . It adapts to the intensity of the current, which is expressed in amps (A).

To make it simple, it is important to know that the larger the diameter of the electric wire, the greater the intensity will be supported .

1.5 mm^2: 10 A

2.5 mm^2: from 16 to 20 A

4 mm^2: 25 A

6 mm^2: 32 A

These data section are representative of those in the housing for a single - phase AC 230 V . Larger sections are necessary to connect the differential meters / switches.

In all power circuits , the grounding is carried out thanks to the insulated stranded wires, which are connected to a grounding pin. This serves to protect people and electrical appliances in case of electric shock or breakdown.

The norm of colors: red, blue and green / yellow

Each type of thread comes in different standard colors:

Red : phase

Blue : neutral

Green / Yellow : Earth (TT)

The phase can also be black or brown in the same installation, but never blue or green / yellow.

TECHNIQUES FOR DECIPHERING THE INDICATIONS OF WIRES AND ELECTRICAL WIRES

Electric cables and wires

The wiring of a house, which covers all the fixed installations, is made up of electrical wires and cables classified by letters and numbers whose function is to indicate their intrinsic characteristics:

03: allowable voltage up to 300 volts. This figure varies depending on the tension;

R: rigid multifiber;

V: polyvinyl chloride (PVC) case. If you have a number in front (2V), it means that it is double (double insulation);

H: harmonized, in accordance with international standards;

K: flexible multifiber;

U: rigid whole core;

G: thread, preceded by the number of those threads and followed by their section. For example 3G2.5 mm² = 3 wires of 2.5 mm².

Example of designation of electric cable

R2V 3G1,5mm²: rigid multifiber, double sheath, admissible voltage 200 V, 3 wires of 1.5 mm².

CODES TO RESPECT ELECTRICAL CABLES

The domestic installations are powered by flexible electrical wires and cables , also classified as H03WF and H05WF .

H: harmonized, in accordance with international standards;

03: 2 wires up to 1 mm²;

05: 2 to 5 wires, depending on the current (mono or three phase), of more than 1 mm²;

W: PVC insulation in outer threads and covers;

F: flexible thread (2F means 2 flexible threads).

H03RTF: transparent cable or coated with a decorative braid (decoration).

H07RNF: Neoprene sheath, wire insulation with elastomer.

Low voltage (BT)

As the name implies, BT is the low voltage , that is the voltage between 0 and 50 V of alternating current (from 0 to 120 V of direct current), so that the low voltage is between 50 and 999 V The BT wire is presented with a copper core covered by an insulator .

The BT is used, mainly, in bathrooms, where the capacities are regulated by the Low Voltage Electro-technical Regulation (REBT) due to humidity and electrification risk.

Purchase unit for electric wires and cables

Electric wires and cables are sold by rolls or reels of:

5 to 10 m;

10 to 25 m;

25 to 50 m;

100 m

Prewired corrugated pipes: an improved solution for recessed installations

CORRUGATED PIPES

Corrugated pipes are plastic pipes that we find in recessed electrical installations, where there are electrical cables. However, corrugated pipes can be delivered with cable ties (the electric wires are inserted into the tube and removed with the cable tie) or prewired (with electric cables), which saves time and easy installation.

With or without wire, corrugated pipes are offered:

with a diameter of 16 to 32 mm;

in coils from 5 to 100 m.

Electrical cables or wires adapted to multimedia, communication and domestic installations

MULTIMEDIA INSTALLATION

The connection for the distribution of DTT signal and antenna wiring is made through the cables formed by a double braided shield with aluminum tape insulation under a protective sheath:

VATCA type 17 or 21 coaxial cable (indicates signal losses. The higher the number that precedes, the better the transmission) The connections can be classified as male or female and can range between 9.50 and 9.52 mm ;

for outdoor use it is preferable to use the 17PATCA coaxial wire , since it is more resistant to UV rays ;

For the digital signal , it is preferable to use DTT cables.

Domestic and communication facilities

Cables for home and communication facilities

The rest of domestic installations are also made with wires, cables and specific connections.

The hi - fi equipment are usually connected by wires 0.75 mm ^{2} red and black (03VH). However, they can be connected for 2 x 1.5 mm² (03VH-H), or even 2.5 mm² or more. The designation OFC ensures a better signal (resistance to oxidation).

For audio / video equipment , connections called RCA are used , which integrate cables with the same name (thermoformed connector).

Home automation equipment, computer networks and communications are connected by RJ45 plugs . Pliers are necessary for these connections.

It is important to respect the standards and recommended sections at the risk of damaging the facilities or causing a fire.

A power supply converts the mains voltage into a lower voltage, which is 220 V to 12 V, 14 V, etc. They can be standard, switched or for electrical panels and to know them we must understand power, voltage, current, types of plug and sockets. Let's see what this is about.

CHAPTER NINE

WHAT IS A POWER SUPPLY

A power supply converts the voltage of a 220 V electrical network into a weaker continuous voltage, between 10 and 30 V, and that adapts better to the electronic devices that equip our home.

There are different technologies but the principle of operation is always the same: adapt the voltage to the device trying to make it as stable as possible regardless of the power required by the device and despite the variations in the network that feeds it.

The output voltage as well as the maximum intensity is two essential characteristics of a power supply.

A good supply is able to maintain the output voltage at its nominal value for all the intensity less than or equal to the maximum intensity of the supply together with the highest possible performance.

A high performance means that a maximum power is restored to the power outlet, which in turn does not heat much. It is good for the environment and also for your pocket!

TYPES OF POWER SUPPLIES

STANDARD POWER SUPPLIES

It is the oldest technology and is also known as linear feeding. A transformer lowers the voltage of the alternative power grid from 220 V to 50 Hz at a weaker voltage (a few tens of volts, but always at 50 Hz).

A first floor called "rectifier "is responsible for transforming this sometimes positive and sometimes negative energy into a voltage that is always positive. Next, the assembly is uniform and the tension becomes greater than or equal to a threshold. A final electronic floor, called regulation, keeps the output voltage constant.

These feeds are rather heavy, expensive and have an average yield ranging from 25% to 50%. The power consumed , and therefore paid, can be between two to four times greater than what the device needs. This type of power supply is ideal for small electrical appliances that generate low consumption. The losses in these cases are not significant and the feeds are compact. Standard power supplies are also often used for devices that are hyper sensitive to electromagnetic disturbances.

SWITCHING POWER SUPPLIES

The switching power supplies , arising from the 70, are both compact, inexpensive and effective time with yield ranging from 75% to 90%. Its own characteristics show its great success in the market.

Its principle is relatively simple. The sector voltage is transformed into a high continuous voltage, around 300 V. A transistor cuts this voltage by opening and closing thousands of times per second.

Then, this set of stimuli is re-transformed into a continuous tension, which is what interests us. The duration during which the capacitor is crossed, allows to regulate the level of output voltage. This voltage must be electronically regulated to be as stable as possible.

These switched supplies have a tendency to create electromagnetic disturbances on the power grid, which can damage other devices. Therefore it is necessary to choose high-end power supplies with effective filters that protect the network from possible disturbances.

LABORATORY FEEDING

As the name implies, it is a type of power that is used in laboratories to power the electronic circuits to experience them and thus make tests on the voltage ranges and the varied intensities.

A laboratory supply allows one or more outputs to be placed on which a voltage ranging from 0 V to a few tens of volts can be regulated. The maximum intensity can also be regulated. Also, stable voltages can be found to power different transistor technologies (2.7 V for TTL technology, 5.5 V for CMOS transistors, etc.).

A laboratory supply has measuring instruments to monitor the voltage and intensity generated in real time. The power supplies Laboratory They are protected against circuit breakers . Even in case of a bad assembly, you will never see this type of food on fire.

POWER SUPPLY FOR ELECTRICAL PANELS

Many appliances and domestic circuits need a low voltage supply , which usually does not exceed 20 V. If these devices are intended to be connected to the electrical panel, then you must opt for a supply that integrates directly into it.

While this solution is somewhat more expensive, it also provides good integration. You should consult your supplier catalog to find the transformer safely for an intercom or communication box in your box .

HOW TO MEASURE A POWER SUPPLY

POWER

The power P generated by the power, indicated in watts, is calculated simply by multiplying the product of the voltage U, in volts, by the intensity I, in amps. That is: $P = U \times I$.

The result will be the maximum power that the block can supply . It is better to choose a higher power, with a 20% safety margin. A maximum power supply heats up and wears prematurely.

It is not uncommon to see that manufacturers of low-end adapters are very optimistic about the effectiveness of their products. Therefore, it is necessary to have a comfortable safety margin.

You must be careful not to exceed the power measurement. You can also buy a food ten times more powerful but you would risk paying more for a block that actually works at a specific performance or whose performance is weak.

TENSION

The adapter must supply the voltage of the powering device. If the voltage supplied is too high , the appliance will burn safely. This implies a risk that is not necessary to run. On the contrary, if the voltage is too weak, the device may not work. If it is slightly weaker, we could try.

It will depend on each case. The cylindrical parts have two connection points with the device: one inside and the other outside. It is necessary to check this polarity . If the polarity is reversed, that is, the - is in the place of + and vice versa, the connected device runs the risk of wear and damage.

Definitely, you must ensure that the voltage and polarity of the device and the power supply are the same.

INTENSITY

As we have just seen, the voltage is set by the device and the power is obtained with the formula $P = U \times I$. The choice of the power of the power supply unit, with a more or less high safety margin, implies the value of maximum power intensity . Preferably, it must be greater than the maximum intensity perceived by the apparatus.

In summary, to choose a power source well:

Determine your tension ;

Calculate its power , function of maximum intensity consumed, with a margin of about 20%.

Without forgetting that:

For weak power, such as battery-powered devices, linear power is recommended;

For strong powers, a switched feed is better thanks to its superior performance;

A good quality feed , although more expensive, better withstand the load and disturb less around it;

If you need a laboratory feed, you should ask about the number and type of output you need. Otherwise, you must determine the type of part that connects the device to its connection power, as well as its polarity in the case of round parts;

If it's a feed to repair, calculate what you need. Do you need different voltages or different output parts?

CHAPTER TEN

SOLAR ENERGY FOR YOUR HOME

If you want to use solar energy in your home, you have several options. You can buy or lease a system or sign an energy purchase contract. The option you choose can affect the amount of money you spend in advance and during the life of the system, the possibility of obtaining certain tax breaks and your obligations when selling your home. Before assuming a commitment, evaluate the company, the product, the costs and their obligations.

SOLAR POWER OPTIONS

If you use a solar panel system you buy less energy from the power supply company and enjoy the benefits of renewable energy.

The Department of Energy says that most houses with solar panels get at least 40% of the energy from the solar energy system; That percentage varies depending on each house. The possibility of solar energy satisfying all your energy needs will depend on the amount of energy your system produces and how much energy you consume.

Buying the solar panel system, may qualify for tax credits or other financial incentives that offset the initial cost. If you lease the system or have a power purchase contract (PPA), you can pay less in advance and the monthly fees may be lower, but you usually cannot receive tax credits or other incentives - as they will be obtained by the company that owns the system.

Even if you buy or lease a system, or have an energy purchase contract, you will probably have to continue paying a portion of the energy you consume to the local power supply company.

HOW TO BUY A SOLAR ENERGY SYSTEM

If you love to buy a solar panel, it is important to pay the fees allotted to such systems but If you are looking for a loan, ask the following:

- How much will you pay in advance?
- What is the percentage rate that will be applied annually?
- How are your payments calculated?
- Will the amount of fees change over the duration of the loan?
- Will you have to make a lump sum payment?

- Can the lender establish a pledge in guarantee on his house or system?
- Incentives and benefits

If you buy a system, you may qualify for credit for federal, state or local taxes or other incentives. The credit for the federal renewable energy tax for homeowners is equal to 30% of the cost of the system. This credit will expire at the end of 2016. The Department of Energy has information on specific state incentives for the use of renewable energy.

You may also receive other benefits for the installation of a solar energy system. Depending on the local net measurement rules, your power supply company may pay you the amount of energy your system returns to the grid.

You may also be able to sell the extra electricity produced by your system or obtain credit for renewable energy certificates (REC). A renewable energy certificate is independent of the amount of electricity produced; It is a certificate stating that you generated a certain amount of renewable energy.

When a business, including a business operating in a residential home, which has solar panels sells all renewable energy certificates, it loses the right to tell its customers that it is using renewable energy. It is important that you consider it if you operate a business from home and want to claim that you use renewable energy.

ANALYZE THE PROPOSALS

Compare the detailed proposals of several companies. The proposals should contain the specific details of the system, namely:

- The expected performance of the equipment and the size of the panels.

The total cost of installation, including all charges for construction permits or necessary electrical work.

- If the production of a certain amount of energy is guaranteed.

What are the guarantees applicable to the equipment (such as panels and power alternators) and to the labor of the installers.

If you own a solar energy system, you have to maintain the panels and equipment - or pay someone to do the maintenance work - unless the seller includes it in the contract.

Maintenance could include the repair or replacement of the power alternator, or the occasional cleaning of the panels in case of low rainfall. Your equipment may be covered by the manufacturer's warranty for an initial period.

THE COMPANY

When you are looking for a company, ask for references to friends, family and neighbors. Check the background of a company at the corresponding state or local consumer protection agencies and before state boards that issue contractor licenses.

Ask if the company you are considering has the licenses, certificates or guarantees required by the authorities of your state, county or city of residence. For example, your state may require the installer to have an electrician's license.

Also do a search on the internet by entering the name of the company and check what you find.

ENERGY PURCHASE CONTRACT

With the purchase power contract a company installs a system in your home and you sign a contract to buy the energy produced by that system. The contracts are long term, and can last 20 years.

Unlike what happens with the lease, you don't pay to use the system, and you don't automatically get all the energy the system produces. You only pay the amount of energy you consume, at a price set by the supplier of the energy purchase contract. Some energy purchase contract providers say they charge a reduced fee for energy because they get tax credits and incentives.

If you rent a system or have a power purchase contract:

ANALYZE THE PROPOSALS

A company could show you a comparison of what you could pay for energy over the next few years with and without the system. In this comparison you can calculate the annual increase that the rates of the electricity supply company will have, and they could suggest that you will pay less if you use your system because you will buy less energy from the electricity supply company. But as the electricity supply company's energy rates depend on several factors, it is difficult to predict future rates.

COSTS, INCLUDING INSTALLATION FEES AND MONTHLY PAYMENTS

The minimum amount of energy a system will produce, and what will happen if the system does not produce that amount.

What will happen if there is a power outage that affects the system installed on the roof?

The guarantees and repairs included and their validity.

What will happen if you need to repair your roof after installing the system?

Read the contract

Before choosing a company, read the contract. Check that the terms of the contract match what the ads and proposals say and what sellers told you. Be clear about the following:

Duration of the contract

What you will have to pay per month (with a lease) or per kilowatt hour (with an energy purchase contract or PPA).

If you will increase the amount of your payments throughout the contract, if so, find out when and how much they will increase.

If you have to pay some other costs or charges.

If the contract includes a "performance guarantee" and how the company will pay you in case the system do not produce the minimum amount of energy.

Who will provide the maintenance and repair service, and if any charges will apply for those services, In the contract you must also say:

Who will receive tax credits or other incentives related to the system.

Who will keep the renewable energy certificates generated by the system.

What should you do to maintain the validity of the contract, for example, pay your bill on a certain date or notify the company if you have plans to sell your home.

What will happen if you want to end the contract early? Will you be charged for early termination or any other charges?

What will happen to the system when the contract ends? Can you renew your lease or energy purchase contract? Can you buy the system? Uninstall it? What will be the price of this?

If you sell your house

Find out if the contract will have any effect on the possibility of selling your home. The provisions of the contract:

Are you allowed to move the system to your new home? How much will it cost?

Are you allowed to transfer the contract to your home buyer?

Do they require you to send a written notice to the company if you want to transfer the contract to your home buyer?

Do they require the buyer to meet the credit requirements or pay some charges before assuming the contract?

CHAPTER ELEVEN

COMMON MISTAKES WHEN INSTALLING SOLAR PANELS

The installation of solar panels, rather than being a trending topic, is an invention that is helping the planet, and our pockets. Investing in a system of solar photovoltaic panels at home contributes to maintaining a great saving in the receipt of CFE.

The interest in the subject, of clean energy, is growing every day, and doubts always arise, questions when looking for an installer or supplier of solar panels, that is why it is important that you take into account these common mistakes when buying panels solar so you don't make them when you're in that process.

SOLAR PANELS HAVE VERY HIGH PRICES

People think that the panels are highly expensive; however, they have to be seen as an investment, not as an expense.

There was a time when the panels were too expensive and it is a date that people still see it as a luxury purchase, but they must consider that having solar panels at home contributes to generating an impressive saving on the CFE electricity bill.

It has always been considered a serious investment for companies and institutions that were looking for savings, even residential buyers preferred not to know the subject because they thought it was quite expensive.

But the cost of solar panel systems has been declining in recent years, and in the same way their performance and efficiency has improved in an indescribable way. Apart from that the government and other institutions have shown funding support to promote the use of solar panels both in companies and in homes.

Thanks to this there are already many companies and homes that save money by taking care of the environment by having a system of photovoltaic solar panels.

When you go to make a purchase of solar panels, the first thing that your installer has to do is generate a quote, this is based on your current electricity bill, and the panels that would be installed, from here they can give you the performance and savings What would you have if you installed the solar panels.

ALL SOLAR PANELS ARE THE SAME

There are thousands of brands and types of solar panels around the world, and it is quite difficult to tell the difference between various types of solar panels. Typically, companies and companies offer the same information about solar panels; starting with the 10 year product warranty, then the 25 year performance guarantee and a nominal power, which is usually around 250 WP.

Therefore the best way to say if a solar panel is good or not, is to ask for references in other countries, and check if there are tests on those solar panels. If there is evidence that certain solar panels are used in large projects in various markets, it means that the solar panels will work perfectly.

THE COUNTRY OF ORIGIN AFFECTS THE QUALITY OF SOLAR PANELS

There is a common ideology between companies and consumers that German-engineered solar panels are better than others.

However, judging a product by its country of origin may become only a strategy to convince the consumer through ideologies already raised with other products, and emotionally motivate them to achieve a purchase of solar panels.

The photovoltaic industry grew exponentially in recent years, therefore some manufacturers could barely adapt, others had to see beyond and had to open facilities in other countries, all this to grow the production of solar panels.

Therefore, it is not fair to assume that the quality of a solar panel has something to do with the country from which it is being manufactured. For example, It is believed that a solar panel manufactured in China is less efficient than one manufactured in a non-Asian market.

However, more than 90% of solar panels imported into Australia in 2013 were made in China. With such an impressive market penetration level, it is hard to believe that Chinese solar panels are of inferior quality.

TECHNOLOGY IS UNSTABLE

Another common misconception about solar panels is that the technology is fairly new, and it is better to wait a few generations before buying them. What most people don't know is that the technology behind photovoltaic modules has been around for more than a century.

While it is fair to say that unstable photovoltaic systems for a long time, recent events have increased the performance of solar panels. And, with the growing interest in clean energy, there is no better time to invest in solar energy than now.

SOLAR PANELS ONLY WORK IF THERE IS ENOUGH SUN

It is a fallacy that solar power only works in the sun. If your business is in an area where there is not much sunlight, then it is not worth investing in electricity.

But, here is the truth. While it is true that most solar modules are more efficient in the sunnier weather, solar panels can still generate a considerable amount of energy even in cloudy conditions.

You have to keep in mind that most photovoltaic systems are installed in addition to an existing power supply in the network. Therefore, a rainy day will not mean that your company will remain without electricity.

Just look at Germany, a country with a fairly rainy climate - they are a solar superpower, getting most of its power from photovoltaic systems. Apart from that the installers must analyze the area where they are going to be accommodated, and they must give the best positioning to the panels so that they can generate a good amount of energy.

IT'S HARD TO FIND A RELIABLE INSTALLER

With such an overwhelming list of manufacturers and installers, it can be difficult to say the good of the less professional. Apart from the generic information, such as "we have installed 5,000 systems", it is difficult for consumers to assess the quality of the service experience they are buying.

As a general rule, when you decide on your solar panel supplier, ask for references, preferably from the companies that have already attended.

The decision to install a photovoltaic solar system is one of the best that will bring numerous benefits. But, before buying photovoltaic modules, research and find an installer that can meet your specific needs.

FAQ FREQUENT QUESTIONS

THE FINANCIAL BENEFITS OF SOLAR ENERGY

By installing a solar energy system on your property, you save money on your electricity bills and protect yourself against rising electricity rates in the future. How much you can save depends on utility rates and solar energy policies in your area.

DOES SOLAR ENERGY HAVE ENVIRONMENTAL BENEFIT?

Solar energy, like other renewable energy resources, has many environmental and health benefits. Using solar and renewable energy, reduces greenhouse gas emissions, which contribute to climate change, and also results in less air pollutants such as sulfur dioxide, which can cause health problems.

WHAT IS NET OR "NETTING" MEASUREMENT?

The net measurement is the system used by public services to accredit the owners of solar energy systems of the electricity produced by their solar panels. With the net measurement, you only pay for the electricity you use beyond what your solar panels can generate.

When installing a photovoltaic solar system in your company or home, the distribution company in your area will install a bi-directional meter to measure the energy entering and leaving your home / business.

HOW DOES USING SOLAR AFFECT THE VALUE OF MY PROPERTY?

If your home has solar power systems it will sell faster and higher in price than a home in the neighborhood without the solar powered system.

However, the value of your property will only increase if you own your own solar panel system. Installing solar for self-consumption will value your property more than the renovation of the kitchen.

HOW DO PHOTOVOLTAIC (PV) SOLAR PANELS WORK?

Energies from the sun is absorbed daily by the Solar panels, these are converted into direct current (DC) electricity. Most homes and businesses operate with alternating current (AC) electricity, so DC electricity is passed through an inverter to convert it into usable AC electricity. At that point, you use the electricity in your home or send it back to the power grid.

HOW MUCH IS THE LIFE AND GUARANTEES OF A PHOTOVOLTAIC SOLAR PANEL?

The lifespan of a photovoltaic solar panel is more than 25 years. Manufacturers' warranties generally cover 10 years for the equipment. The production of these panels is guaranteed for 25 years, based on specific data for each year (varies by manufacturer).

DO MY SOLAR PANELS PRODUCE ENERGY WHEN THE SUN DOES NOT SHINE?

The amount of energy that your solar energy system can generate depends on the intensity of sunlight. As a result, its solar panels will produce a little less energy when the sky is cloudy, and there is no energy at night. However, due to high electricity costs, solar is a smart decision, even if you live in a city with cloudy weather.

CAN I LEAVE THE NETWORK WITH SOLAR PANELS?

Provided the system was installed for private consumption, you will still be connected to the network. This allows you to draw energy from the network when your system is not producing all the power you need and return the energy to the network when it produces more than it uses.

It is possible to leave the network with a solar energy system that includes battery storage, but it will cost more and is unnecessary for most owners. Only if you are in an off-grid site we recommend the use of solar with batteries.

WILL I STILL RECEIVE AN ELECTRICITY BILL IF I HAVE SOLAR PANELS?

If you do not want the electricity bill then you will have to go completely off-grid, you will continue to receive a bill from your utility company.

However, you can observe a drastic reduction in your bill, or even reduce the amount billed to almost zero, this is a tailored suit, so your solar panel system must adapt to your energy use.

WHAT IS THE SITUATION OF THE SYSTEM DURNG BLACKOUT?

If your solar panel system is connected to the network, it will shut down in case of a blackout. This is to prevent people from the maintenance service teams of the electric companies or emergency services from being injured by the panels that send the electricity to the grid.

However, there are certain investors that you can buy that provide backup power in a power outage in case it is tied to a battery. You can also install a panel of critical loads.

WHAT TYPE OF SOLAR PANELS SHOULD I USE, POLYCRYSTALLINE OR MONOCRYSTALLINE?

This is answered based on the availability of space, mono crystalline normally generate more energy per square meter than polycrystalline. Then, depending on the needs of the project, it is appropriate to evaluate what is most appropriate, considering future requirements, budget and available space.

MICROINVERTERS OR "STRING" TYPE INVERTERS?

The difference between these is that micro inverters transform solar energy into each of the solar panels from direct current to alternating current, and "String" type inverters do so in groups of panels.

Therefore, if a solar panel of a specific group is affected in its production, it will affect the production of the group. Our choice will be dependant on the weather conditions of the installation site, type of roof and the size of the solar plant. Although the trend in many countries is towards micro investors due to security regulations since the cost / benefit ratio is getting better.

IS MY ROOF SUITABLE FOR SOLAR PANELS?

South-facing roofs with little or no shade and enough space to fit with a solar panel system are ideal for solar installation. However, in many cases there are solutions if your house does not have the ideal sunroof.

All installation offers are based on the actual conditions of your roof. Important: the orientation to the South or North of the solar panels will depend on the latitude in which your home or company is located.

Latitude north, the panels should look south (and vice versa).

WHAT IS THE APPRIOPRATE SYSTEM SIZE?

The overall size of your solar system will depend on the amount of electricity you use monthly, as well as the weather conditions in which you live. Always start by investing in energy efficiency, then solar.

DO I NEED TO REPLACE MY ROOF BEFORE INSTALLING SOLAR ENERGY?

Solar energy systems can last 25 to 35 years, and it can be expensive to remove and reinstall them if you need to replace your roof. If your roof needs maintenance in the short term, you must complete it before finishing your solar installation.

We can recommend the most appropriate for your case. In some cases it may be partially replaced or the cost of reinstalling the solar panels will be small. Remember in those cases only work with solar panels. Solar panels are only part of the photovoltaic solar system.

DURATION OF THE SOLAR ENERGY

Solar panels are very durable and able to withstand snow, storm showers, wind and hail. The different components of your solar energy system will have to be replaced at different times, but your system must continue to generate electricity for 25 to 35 years.

HOW MUCH HAS THE PRICE OF SOLAR ENERGY FOR RESIDENTIAL USE DECLINED IN RECENT YEARS

If you are an optimist looking for statistics to feel good, the cost of solar electricity in the last decade is a great place to start. The cost of solar installation has been reduced by about 70 percent in the last 10 years.

The solar market last year experienced a five percent decrease in cost. There is no doubt that solar energy has evolved from a clean technology product to a reasonable home improvement that millions of people are coming on board in 2019.

WHAT IS THE DIFFERENCE BETWEEN SOLAR ENERGY FOR BUSINESSES AND SOLAR ENERGY FOR DOMESTIC USE

A commercial solar project could boost a city or the operations of a company. In comparison, residential solar systems tend to maintain a constant size (6 kilowatts on average).

Thanks to their relatively small scale, the solar panels on the roofs for the home are an attainable energy improvement that can generate significant savings in the electricity bill for the owners of any income level. On the other hand, commercial solar energy requires a large investment and a collective group of investors.

HOW MUCH DO RESIDENTIAL SOLAR PANEL SYSTEMS COST

It is based on the state and size of the system. However, there are data that can help you estimate the cost of solar panels in 2019. The easiest way to calculate the cost of solar electricity in different system sizes is in dollars per watt ($ / W), which indicates how many solar energy dollars Cost per watt of electrical production available.

In 2019, owners pay an average of $ 2.98 / W. To put that figure in perspective, in 2008 the average cost of solar energy was a little more than $ 8 / W. For an average system of 6kW, a price of $ 2.98 / W means that you will pay approximately $ 17,880 before taxes and tax refunds in 2019. (Note: This data belongs to the US each country has its own)

CONNECTING MY SOLAR TO THE NETWORK? WHAT IS NET MEASUREMENT

This will depend on each country. The vast majority of domestic solar systems will be connected to the network. With solar energy connected to the grid, the net measurement serves as an efficient solution to the question "How am I going to power my solar home at night?"

The net measurement is a solar incentive where you receive credits when the solar system produces Too much electricity At times when your panels do not produce enough electricity, you can use those bill credits to cover the cost of using your network electricity.

If you are not connected to the network, you will not have access to your utility's electricity. This means that, to build a completely off-grid project, you will need extensive energy storage capabilities, an extra-large solar panel system and backup power arrangements to cover it when your panels do not receive enough sun.

INSTALLATION TIME PERIOD FOR SOLAR ENERGY

if you are configuring the net measurement, that process will increase the additional time until your panels are correctly connected to the network. In general, although the decision process of solar panels can take some time, the installation time is very fast and quite simple.

IMPORTANCE OF ROOF QUALITY AN SOLAR PANEL

Provided the residential solar sector is the list of options for homeowners who want to use solar energy but do not have a suitable roof. Solar floor-mount installations and community solar gardens are two common ways to access solar energy without installing anything on the roof.

Community solar energy involves connecting with members of a group or your neighborhood to share a solar system, while ground-mounted arrangements are an easy way to own and install your own system without going through the roof obstacles.

DOES SOLAR ENERGY MAKE SENSE IF I DON'T PLAN ON BEING IN MY HOME FOR 25 YEARS

A common concern for homeowners considering solar energy is: "What happens if I move after installing solar panels?" A typical solar panel system lasts 25 to 30 years. If you do not plan to own your home for so long, you may wonder if solar energy still makes sense.

The good news is that solar energy increases the value of your property and can accelerate the process of selling the property when the time comes. The housing market is full of buyers excited by the prospect of acquiring a solar house that comes with the benefit of zero utility bills.

WHAT PERCENTAGE OF YOUR HOME CAN BE POWERED BY SOLAR

solar panel system can theoretically compensate for all its energy use, it is not realistic to expect that level of panel production every day of the week.

The leading solar power manufacturer in the United States, Sun Power, recommends homeowners a factor of

25 percent when calculating their goal for solar panel compensation. Solar panels cannot work with maximum efficiency all the time. There will be certain days when the connection to the network is necessary to completely cover your energy use. But nevertheless,

WHEN WILL YOUR DOMESTIC SOLAR SYSTEM REACH THE "BREAK-EVEN POINT"

Many owners are very interested in calculating the recovery period of their solar panel, which is the amount of time it will take to save the electric bill to offset the cost of installing the solar panel. The expected balance point on average for homeowners in the US UU, It is approximately 7.5 years old.

CONCLUSION

Solar energy remains one of the best and comfortable ways to power our homes. Despite the inconveniences, the use of solar energy has increased by about 20 percent per year in the last 15 years, thanks to the rapid fall in prices and the increase in efficiency.

Japan, Germany and the United States are important markets for solar cells. With tax incentives and efficient coordination with energy companies, solar electricity can often pay for itself within five to ten years.

It is an excellent alternative to cover and complement the energy needs of small cities or homes, today it is impractical when using this technology as the main source of household energy, but it is a great option to use as complement.

If you are a person committed to the environment, this type of renewable energy is undoubtedly the best option to take care of the environment; it is clean renewable and does not harm the environment.